STRUCTURE AND SUCCESSION OF FOULING COMMUNITIES

OCEANOGRAPHY AND OCEAN ENGINEERING

Additional books in this series can be found on Nova's website at:

https://www.novapublishers.com/catalog/index.php?cPath=23_29&seriesp=Oceanography+and+Ocean
+Engineering

OCEANOGRAPHY AND OCEAN ENGINEERING

STRUCTURE AND SUCCESSION OF FOULING COMMUNITIES

V. V. KHALAMAN

Nova Science Publishers, Inc.

New York

LIBRARY OF CONGRESS CATALOGING-IN-PUBLICATION DATA

Available Upon Request

ISBN: 978-1-60876-827-1

Published by Nova Science Publishers, Inc. ✝ *New York*

CONTENTS

Summary   **vii**

Introduction   **ix**

Chapter 1   Development of Fouling Communities   **1**

Chapter 2   Life Strategies. New Approach to an Old Problem   **5**

Chapter 3   Development of Fouling Communities in the White Sea   **9**

Chapter 4   Materials and Methods   **11**

Chapter 5   Results and Discussion   **13**

Acknowledgments   **33**

References   **35**

Index   **49**

SUMMARY

The development of fouling communities in the White Sea cannot be described as a linear sequence of stages replacing each other, but looks like a network of possible stages. Such scheme has a low predictive value and does not explain changes occurring in fouling communities. The estimation of a survival strategy of epibenthic organisms in terms of Ramenskii-Grime's system (Ramenskii, 1935; Grime, 1974) is conceived as the most promising method in analysis of structure and development of fouling communities. Testing of this hypothesis was carried out in five fouling organisms: mollusks *Mytilus edulis* and *Hiatella arctica*, solitary ascidians *Molgula citrina* and *Styela rustica*, and sponge *Halichondria panicea*. Characterization of fouling organisms was based on autecological traits of testing species. *M. edulis* and *S. rustica* adopted a competitive strategy. *M. citrina* is a ruderal organism. *H. arctica* exhibited a tolerance strategy, but *H. panicea* demonstrated unusual mixed competitive-ruderal one. The obtained strategies well coincide with real positions of testing species in structure and succession of fouling communities developing in the White Sea.

INTRODUCTION

The communities of sessile organisms that grow on solid substrates of both natural and artificial origin have been the subject of numerous studies for over one hundred years. The reasons for this are quite obvious. The principal reason lies in the adverse impact caused by the fouling organisms that inhabit ship bottoms, marine oil platforms, piers and other artificial objects. At the same time, the techniques used for culturing some aquatic invertebrates, particularly the oysters and mussels, are largely based on the ability of these organisms to form fouling communities. Apart from their purely utilitarian value, the fouling communities have become a very convenient model for studying a range of theoretical problems of marine biology, such as structure, resistance and development of epibenthic communities, and competition among marine organisms (Anger, 1978; Dean, Hurd, 1980; Dean, 1981; Barkai, Branch, 1988; Myers, 1990; Glasby, 2001; Rajbanshi R., Pederson J., 2007 and many others).

Chapter 1

DEVELOPMENT OF FOULING COMMUNITIES

There is still no single unifying view on the development of the fouling communities, despite the long and rich history of studying this problem. The existing diversity of ideas could be reduced to two different viewpoints.

The first of these two viewpoints regards the development of biofouling as succession, i.e., the replacement of certain communities by others until a climax state is reached. The main criterion is a recurrence of these processes and low dependence on external factors. These successions have been described many times, mostly, for the seas in the boreal and subtropical zones (Scheer, 1945; Millard, 1951; Reish, 1964a; Haderlie, 1969; Brayko, 1974; Saenger et al., 1979; Gorin, 1980; Rudyakova, 1981; Chalmer, 1982; Oshurkov, 1985; Okamura, 1986b; Hirata, 1987; Vance, 1988; Henschel et al., 1990; Khalaman, 2001b).

The proponents of the alternative viewpoint state that the development of fouling communities cannot be described as succession (Sutherland, 1974; 1978, 1981; Sutherland, Karlson, 1977). They argue that there is no single climax, that the realization of any given community depends on a number of poorly predictable factors and that the development of fouling in general is essentially stochastic in nature (Kajihara et al., 1976; Osman, 1977; Anger, 1978; Breitburg, 1985; Rao, Balaji, 1994).

It should be noted here that there have been only a few studies involving sufficiently long-term monitoring of the fouling communities. However, all these studies indicate that the development of epibenthic communities is a very lengthy process, ranging in duration from some years to several decades (Gulliksen et al., 1980; Wendt et al., 1989; Oshurkov, 1992; Oshurkov, Ivanjushina, 1992; Sell, 1992; Butler, Connolly, 1999; Whomerslay, Picken, 2003; Khalaman, 2005a). It was also observed that the rate of changes in the fouling communities decreases as they grow older (Butler, Connolly, 1996).

The majority of studies, dedicated to the development of fouling communities, are relatively short-term (usually one or two years in duration) series of observations dealing with developing fouling assemblages on newly installed experimental plates. In this case, the investigators have the opportunity to study only the early stages of development of fouling communities, usually dominated by short-lived species. The formation of these fouling communities is greatly influenced by seasonality, which to a large extent explains their high variability and stochastic mode of development (McDougall, 1943; Brajko, Dolgopol'skaja, 1974; Sutherland, Karlson, 1977; Kawahara et al., 1979; Riggio, 1979; Turpaeva, 1987; Bailey-Brock, 1989; Todd, Turner, 1989; Turner, Todd, 1993; Underwood, Anderson, 1994; Nandakumar, 1996; Brown, Swearingen, 1998 and many others). The succession, as a process determined by interspecific interactions (Shelford, 1930; Connell, Slatyer, 1977; Dean, Hurd, 1980), could be obscured and even inhibited by seasonal changes (Calder, Brechmer, 1967; Mook, 1981a; Greene, Schoener, 1982; Shim, Jurng, 1987; Rajagopal et al., 1997).

The life spans of the organisms that are responsible for seasonal changes of the fouling communities as a rule do not exceed 1-1.5 year. When an assemblage of fouling organisms becomes dominated by long-lived species, the seasonality ceases to play the guiding role in the further development of this community. Therefore, the results obtained in the short-term studies cannot be generalized as an accurate reflection of the developmental patterns of the fouling community as a whole.

There are cases, when, for some reason, the long-lived species are absent or cannot exist in those conditions, under which the fouling community develops. As a rule, this is explained by periodic disturbances, associated with storms (Sutherland, 1974, 1975; Russell, 1975; Sutherland, Karlson, 1977), or the action of monsoon rains (Venkat et al., 1995) or ice (Bowden et al., 2006). In some cases, the long-lived organisms "slough off" from the substrates due to the death of the earlier colonizing organisms that reside underneath (Saenger et al., 1979). The development of fouling assemblages could be significantly influenced by predation (Mook, 1981a; Barkai, Branch, 1988; Okamura, 1986; Fitzhardinge, Bailey-Brock, 1989; Osman, Whitlatch, 2004). The surfaces denuded by predators become colonized again by short-lived organisms. The succession as a whole acquires a cyclical pattern, and the development of the fouling community becomes limited by the existence of short-lived organisms. The period when these species predominate in the community could be named the short-lived organism phase. This phase is quite variable, and a type of fouling community that is determined by dominant species is greatly influenced by seasonality.

When the long-lived species attain dominance in the fouling communities (in the beginning of the long-lived organism phase), the further development of fouling is not always strictly determined. In the boreal zone, the long-term communities are most often dominated by the bivalves: mussels or oysters (Scheer, 1945; Reish, 1964a; Brayko, Dolgopol'skaja, 1974; Oshurkov, 1985; Hirata, 1987; Ardizzone et al., 1989). However, ascidians, sponges, and soft corals could also become dominant species (Henschel et al., 1990; Oshurkov, 1992; Sell, 1992; Butler, Connolly, 1999; Khalaman, 2001a). Moreover, the long-term fouling assemblages that develop in similar conditions could be dominated by different species. The situation, described for the White Sea (Oshurkov, 1992; Khalaman, 2001b, 2005), could be taken as an example. The long-term fouling assemblages living there at depths of 1 to 5 m are dominated either by mussels (*Mytilus edulis*) or by solitary ascidians (*Styela rustica*). The settlements of these species are successively replacing one another. This replacement occurs according to inhibition model of succession, due to the lack of timely recruitment in settlement of the replaced species (Khalaman, 2005a).

A diversity of development scenarios for the fouling communities has led a number of authors to search for a certain common property, shared by any given replacement series. Such property was found in a replacement series of dominant forms leading from *r* to *K* strategists, i.e. the substitution of more opportunistic species by less opportunistic, and short-lived by long-lived (Luckens, 1976; Anger, 1978; Kay, Butler, 1983; Rocha, 1991; Hirata, 1992; Whomerslay, Picken, 2003; Khalaman, 2005a). The stochastic model for the development of epifaunal communities ("fixed lottery"), based on the principles of Markov model (Green, Schoener, 1982), predicts a community "shift" toward increasingly more long-lived organisms, and Glasser (1982) points out explicitly that succession should always end with long-lived species, regardless of how (when) it actually began, provided that no destruction in biological or physical factors occurs.

Consequently, if there are no disturbances caused by extrinsic factors, the only strictly regular pattern in the development of fouling communities is the sequence of phases: the "short-lived organism phase" followed by the "long-lived organism phase." If the roles of both the short-lived and the long-lived organisms, under certain conditions, are taken only by a single species, the development of fouling is strictly determined. This development of fouling progresses as a traditional succession, in which one dominant species (short-lived organism) is replaced by another (long-lived organism).

However, one role is more often claimed by several species. This is common for the short-lived organism phase, and probably less so for the long-lived organism phase. In this case, the general development scheme for fouling is not a linear sequence, in which one species is replaced by another, but rather a network of possible states. The course that will be followed by the fouling community in each particular case depends on a number of poorly predictable factors.

PROBLEM

Several theories were proposed to explain the development of the fouling that does not fit the traditional concept of succession as a sequential replacement of communities (Clements, 1916): "multiple stable points" (Sutherland, 1974; Sutherland, Karlson, 1977), "stable endpoints" (Osman, 1977), "fixed lottery" (Green, Schoener, 1982) and "alternative states" (see review in Petraitis, Dudgeon, 2004). However, all these concepts describe the overall pattern of development for the fouling communities, rather than answering the following important question: "What determines the position of a given species in the community and in the succession?" No credible prediction of the community formation is possible without having a clear answer to this question. The solution lies in a different approach to studying the succession of fouling communities, specifically, in estimating the life strategy of each individual species, participating in this process.

Chapter 2

LIFE STRATEGIES. NEW APPROACH TO AN OLD PROBLEM

The r/K concept of life strategies has gained wide acceptance in animal ecology. This concept is based on the logistic model for population growth. The pressure of external factors leads to r- and K- selection, and the advantage is gained either by small, short-lived organisms, that invest a substantial amount of resources in reproduction (r-strategists), or by large long-lived species having relatively small reproductive efforts (K-strategists) (Pianka, 1970). It is believed that K-strategists exist in stable living conditions under high population density and at high levels of intraspecific competition. In this case, the population growth is limited, and mortality depends on population density. By contrast, r-strategists are characteristic of the habitats, affected by periodic catastrophic disturbances. Intraspecific competition in these organisms is considerably less intense, the rates of population growth are consistent with exponential model, and mortality is independent of population density (Odum, 1983; Begon et al., 1990).

Ramenskii (1935) has proposed a more flexible system for terrestrial plants. He described three major life strategies: tolerance (S) ("patient" in Ramenskii), ruderal (R) ("explerent" in Ramenskii) and competitive (C) ("violent" in Ramenskii) strategies. This classification was adopted in the Russian literature, but received little attention elsewhere. Almost 40 years later the classification of plant strategies was "rediscovered" by Grime (1974, 1979).

Competitive (C) and ruderal (R) strategists are comparable to K- and r-strategists, respectively. Some authors even consider tolerance (S) as an additional vector for $K - r$ axis (Begon et al., 1990). However, the triangular Ramenskii-Grime system was based on somewhat different principles. In this system, the strategy is determined by a combination of two extrinsic factors: disturbance intensity and stress intensity. The disturbance is defined as a substantial removal of the population biomass by its consumers or a mass mortality of organisms due to the effects of extreme abiotic factors. The stress limits biomass growth and population settlement density by limiting the resources or by affecting suboptimal physical conditions of the environment. A combination of weak disturbances and intensive stress defines tolerance strategy (*S*), strong disturbances and weak stress – ruderal strategy (*R*), and weak disturbances and weak stress – competitive strategy (*C*) (Ramenskii, 1935; Grime, 1974, 1979).

Tolerance strategies were divided into two categories, referring to plants, which are tolerant to unfavorable abiotic environmental conditions and to a high degree of competitive pressure from other species, respectively (Mirkin, Rosenberg 1983). Another classification scheme of tolerance strategies was suggested by Campbell and Grime (1992), who distinguished species following a disturbance (D) – and a stress – tolerance (S) strategy, respectively.

Tolerance, ruderal and competitive strategies could be viewed as some extreme points, that serve as the limiting values for the actual biocoenotic (ecological) properties of a given species. Apparently, the species with "pure" strategies do not exist in nature. Each species possesses a certain combination of tolerance, ruderal and competitive properties, but no species can have all three, exerted to the maximum degree. The characteristics of any organism could be represented by points within the triangle, whose apexes are occupied by the mentioned above strategies (Grime, 1979). The evaluation of the species strategy is relative. The life strategy of the same species in various localities and in various communities could be quite different. It depends on abiotic environmental conditions and species composition, i.e., the properties of the organisms that coexist with the species being evaluated (Mirkin, Rosenberg 1983; Rabotnov, 1993).

TASK

The sessile life style is a common feature of both terrestrial plants and sedentary organisms that constitute the base of fouling communities. Consequently, the Ramenskii-Grime system appears promising as a potential tool for evaluating marine epibenthic animals. The purpose of this paper is to present such an evaluation.

The goal of this study is two-fold: 1. To estimate life strategies of certain fouling species, in Ramenskii-Grime system, by using key biological characteristics of these organisms. 2. To compare obtained characteristics with the actual position of the species in the structure and succession of fouling communities.

The fouling communities that develop in the White Sea in the upper 5 m of the water column were selected as a model for this study. The hydrological characteristics of the White Sea and development patterns of fouling in this water basin should be described in more detail.

STUDY AREA

The White Sea is a small internal marine basin (with an area of approximately 91 000 km^2), which is a part of the Arctic Ocean (Figure 1). It is connected to the adjacent Barents Sea by a relatively narrow strait known as Gorlo (Throat). The mean depth of the sea is about 60 m; the maximum depth is 343 m. The salinity at upper layer of water ranges from 24 to 27‰, and reach as high as 30‰ at greatest depths.

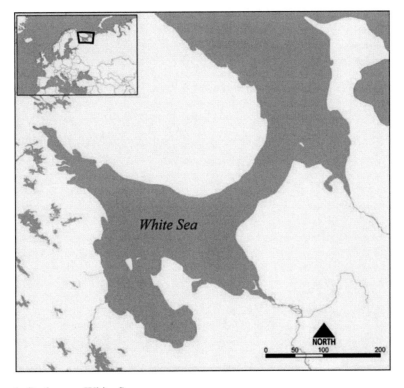

Figure 1. Study area. White Sea.

In spring, during the period of rapid melting of snow and ice, the salinity of the surface water layer (0.5 m depth) in small inlets could fall to 0‰. In winter, the water temperature could drop up to freezing point and could be as low as -1.4 – -1.8°C. The White Sea is covered by ice from November or December to May, although there is usually no solid ice cover in the central region of the sea. In summer, the temperature of the surface water layer in various areas of the White Sea ranges from 6°C to 15°C, and in shallow, upper portions of the bays and small inlets, the water could be warmed up to 20–24°C. Temperature decreases with depth. The water is sufficiently heated in summer only down to 5–15 m. At depths below 50-70 m, the water temperature never rises above 0°C (Babkov, Golikov, 1984; Berger et al., 2001).

Chapter 3

DEVELOPMENT OF FOULING COMMUNITIES IN THE WHITE SEA

The studies of the fouling communities in the White Sea began in 1960s (Zevina, 1963), and the principal developmental patterns of these communities are currently well known (Sirenko et al., 1978; Oshurkov, 1985, 1992, 2000; Maximovich, Morozova, 2000; Khalaman, 2001a, b, 2005a; Railkin, 2004 and many others).

The presence of the summer thermocline affects the development of fouling in the White Sea. In particular, the mussel's community, which is very common in fouling for the boreal seas, develops in the White Sea at depths of less than 5 m. The development of fouling proceeds much slower at greater depths; more cold-tolerant forms, such as ascidians, sponges and bryozoans, tend to predominate there.

On the other hand, ascidians are usually not found above 1 m depth due to significant springtime decrease in surface water layer salinity and due to the summer heating of this layer (Oshurkov, 1985; Khalaman, 2001a).

The development of a fouling community in the upper 5 m of the water column in the White Sea could be presented graphically as a diagram (Figure 2). Apart from microbial film, any exposed surface is first colonized by "short-lived" organisms.

A corresponding phase in the development of the fouling assemblages, depending on the season, hydrological characteristics of the region and physical properties of the substrate, could take one of the following main forms:

- community of filamentous algae and/or hydroids;
- community of bryozoans and sedentary polychaetes of the family Serpulidae;

- community of the barnacles *Balanus crenatus*,
- community of ascidians of the genus *Molgula*.

The "short-lived" organisms are replaced by the "long-lived" organisms that form two principal communities: that of the blue mussel (*Mytilus edulis*) and that of the solitary ascidian *Styela rustica*. Under some circumstances, the mussels could foul directly on a clean surface. In this case, the phase of "short-lived" organisms is eliminated from succession (Khalaman, 1993, 2001b).

The community of *Styela rustica* can exist in several forms depending on the extent to which a substrate is occupied by the brown alga *Laminaria saccharina* and the sponge *Halichondria panicea*. The reason for this is that this type of fouling is an equivalent of the kelp zone epibenthos, which is typical for the upper subtidal zone of the White Sea (Myagkov, 1979; Golikov et al., 1985, 1988; Plotkin et al., 2005). For this reason, *L. saccharina* could even predominate in the fouling community in terms of biomass. However, it is *Styela rustica*, that is the most long-lived, enduring and always abundant component of these fouling communities (Oshurkov, 1992; Maximovich, Morozova, 2000; Khalaman, 2001a). If, however, the substrate, on which a fouling assemblage develops, is prone to periodic drying due to the ebb and flow of the tides, this substrate becomes colonized by a community of brown algae (*Fucus vesiculosus* and *Ascophyllum nodosum*), characteristic of the lower tidal zone of the White Sea (Oshurkov, 1992).

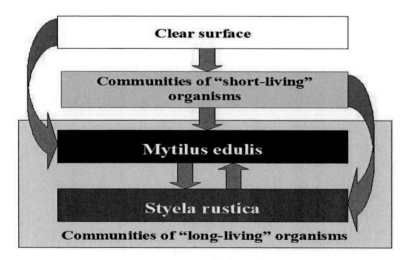

Figure 2. General scheme of development of fouling communities in upper (0-5 m) water layer of White Sea (by Khalaman, 2005a).

Chapter 4

MATERIALS AND METHODS

TESTED SPECIES

Five most common fouling species, that form long-term fouling communities in the upper 5 m of the water column in the White Sea, on the substrates not affected by periodic aerial exposure, were selected for this study. These are: the bivalves *Mytilus edulis* and *Hiatella arctica*, ascidians *Molgula citrina* and *Styela rustica*, and the sponge *Halichondria panicea*. All these species belong to the same life form: sedentary filtrators. However, the mollusks and ascidians are solitary organisms, and the sponge is a modular organism.

USED CHARACTERISTICS

The life strategies of the selected species were evaluated using the following parameters and characters:

1. Definitive size
2. Life span.
3. Growth rate.
4. Fertility and the volume of larval pool.
5. Tolerance to changes in main environmental factors (temperature, salinity).
6. Edificatory abilities of monospecific settlements, i.e. ability to modify environmental conditions and provide other species (associated fauna) with shelter, food, substrate, etc.

7. Attachment strength
8. Pumping rate.
9. Resistance to invasion of other fouling organisms.

The characters selected, on one hand, are relatively easily tested and known for most of the species, and, on the other hand, provide sufficiently complete characteristics for ecological features of the species that belong to this life form.

The species evaluation was carried out separately for each individual parameter. The degree to which the character is manifested in a given tested species was estimated in relation to the same parameter in other tested species. In other words, the estimation was relative and ranked.

According to the degree of character manifestation (strong (+), weak (-) or intermediate degree (+/-)), the species was given one point for the scales of competitor (C), ruderal (R) or tolerance (S) strategies. If a certain character could be attributed to two strategies at once, one additional point was given for each of the strategies. The characters were assigned to strategies in line with the principles used in geobotany. For instance, S-species are characterized by low growth rates and small definitive sizes, but, at the same time, by a high degree of resistance to extreme effects of abiotic and/or biotic extrinsic factors. C-species are characterized by a considerable life span, a relatively high growth rate and large definitive sizes, and have the ability to seize resources in a rapid and efficient manner. R-species are small organisms that have low competitive abilities, but, at the same time, high fertility, considerable growth rates and a short life span (Grime, 1979; Vasilevich, 1987; Rabotnov, 1992, 1993; Romanovskii, 1998).

The total scores for each of the nine characters received by the species were summed up. The final scores obtained by the species for competitor (C), ruderal (R) and tolerance (S) strategies, were used as data, indicating the species location in the C-S-R space, defined by a 3-dimensional Cartesian coordinate system.

Chapter 5

RESULTS AND DISCUSSION

DEFINITIVE SIZE

Mytilus edulis. The mussel *Mytilus edulis* inhabiting the fouling communities of the White Sea could have a shell as long as 80 mm or even slightly longer; but usually the shell length reaches 60-70 mm (Sukhotin, Kulakowski, 1992; Naumov, 2006).

Hiatella arctica. A.D. Naumov has indicated in his survey (Naumov, 2006) that the largest specimen of *Hiatella arctica*, collected in the White Sea, had a shell 33.5 mm long. The largest specimen collected by the author in the fouling communities, was 39 mm long. It should be noted, however, that the specimens of *H. arctica*, reaching 30 mm or more in size, are very rare in a fouling.

Styela rustica. The body dimensions of ascidians of the genus *Styela* are variable. When agitated, they can contract by several times of their original length. In general, however, *Styela rustica,* in both contracted and relaxed states, are comparable in size with mussels. The maximum body height of this animal, in relaxed state, is about 70 mm (Olifan, 1948; Khalaman, in press). To analyze the size structure of the *S. rustica* settlements (Khalaman, 2005a) and growth rates (Khalaman, in press) of these animals, the author has used the L_{comb} parameter, calculated from the following equation:

$$L_{comb} = \sqrt[3]{H * \pi * (D/2)^2}$$

where H is the body height and D is the body width of an ascidian in the state of maximum contraction. This approach was dictated by the necessity to normalize the measures of body condition and compensate the differences in body shape for different specimens. *S. rustica* has more or less cylindrical body shape, but the height-to-width ratio in these animals could vary to a large extent. The largest specimen of *S. rustica* was found by the author in 1996 in the fouling community living on artificial substrates of the mussel culture farm. This specimen had the following parameters: L_{comb} = 28.99 mm, wet weight (W) = 20.75 g.

Molgula citrina. This species has small definitive sizes. The body shape is nearly spherical. The body diameter of adult specimens is 13–16 mm (Redikortsev, 1916; Olifan, 1948). According to the author's own observations, the specimens of *M. citrina*, with the body diameter over 13 mm, are rare in the fouling communities of the White Sea.

Halichondria panicea. The sponge *H. panicea* is a modular animal and, therefore, the parameter selected to be evaluated was the mean colony size. It should be noted that the development of this sponge is seasonal. The life cycle phases of *H. panicea* in the North (Barthel, 1986) and Barents (Ivanova, 1981; Yereskovski, 1994) Seas have somewhat different time frames, but the life cycle itself remains essentially similar. In the Barents Sea, the active growth of the colonies begins in the end of hydrological winter and slows down in May-June, due to the beginning of the spawning period. After the spawning season is over, the sponges begin to degrade, until the maternal specimens become completely disintegrated in October or November. In the White Sea, *H. panicea* spawns in July, but substantial colony degradation in fouling communities of the White Sea is observed during the second half of August or in September. According to the author's own observations made at the White Sea, during the summer months, the areas covered by individual colonies of *H. panicea* in the fouling communities are usually less than 2 dm^2.

The results presented in this section show that the species could be arranged in order of increasing size, as follows: *M. citrina − H. arctica − S. rustica − M. edulis − H. panicea.* The last three species could be regarded as large-sized, the first two as small-sized. Large definitive sizes are characteristic of *C*-species, whereas small sizes could be found in both *R*- and *S*-species.

LIFE SPAN

Mytilus edulis. The life span of this mollusk in the natural biotopes of the White Sea depends on the habitat and is estimated by various authors to be 5 to 17 years, though some specimens could live until 25–26 years of age (Naumov,

2006). In the fouling communities, including mussel cultivation installations, the life span of these mollusks averages 5–9 years (Maximovich et al., 1996).

Hiatella arctica. Various authors give disparate estimates for the life span of this species. For *H. arctica*, living in the North Atlantics, Petersen (1978) estimates it to be 18 years, and Sejr et al. (2002) give an age estimate of 126 years. The latter estimation should probably be dismissed as an odd artifact, resulted from an error in application of acetate peel method.

According to Matveeva and Maksimovich (1977), the maximum life span of *H. arctica* in the White Sea is 6 years. The analysis of the size-frequency structure dynamics for the settlements of this species in the fouling communities suggests that the life span of these animals averages 4–6 years, although some individuals could probably survive until 10 years of age (Khalaman, 1993).

Styela rustica. The analysis of the size-frequency structure dynamics for the settlements of *S. rustica* growing on the mussel cultivation installations in the White Sea suggests that the average life span of these animals is approximately 7 years (Khalaman, 2005a). These results were confirmed through a direct field experiment. The tadpoles of ascidians were placed on the bottom of the plastic Petri dishes in the laboratory. Each settled specimen was labeled. The dishes were then placed in the sea. The ascidians were periodically observed and measured for a period of several years. It was shown that the life span of *S. rustica* is 7 years (Khalaman, in press).

Molgula citrina. There is no accurate data on the life span of these ascidians living in the White Sea. The author's own several years-long observations show that their life span, apparently, does not exceed one year. *M. citrina* spawns in mid-summer (July), and, by the end of this month, this species almost completely disappears from the fouling communities, because most specimens die after the spawning.

The juveniles of *M. citrine* appear in great numbers in August and attain adult sizes as early as September or October of the same year. *M. citrina* does not seem to differ from other species of *Molgula* in terms of life span (Frame, McCann, 1971; Sutherland, 1974).

Halichondria panicea. As mentioned earlier, the development of the colonies of *H. panicea* is seasonal. Their life span could be estimated to be 6-7 months.

To sum up, *H. panicea* and *M. citrina* are short-lived species and could be classified as *R*-strategists. *M. edulis*, *S. rustica* and *H. arctica* are long-lived species that have more or less equal life span. The long life span is characteristic of both *C*-species and *S*-species.

GROWTH RATE

This parameter is associated with two parameters described in the previous sections: life span and definitive sizes.

Halichondria panicea and *Molgula citrina*. It is fairly obvious that *H. panicea*, whose colonies attain maximum size by mid-summer, is characterized by the high growth rates. There is no accurate data on the growth rates of *H. panicea* living in the White Sea. It is known, however, that the same sponges living along the shores of Denmark, grow by 2.8% to 4% a day (Thomassen, Riisgård, 1995).

M. citrina could also be regarded as rapidly growing organisms, because juveniles of these ascidians emerging in July through August attain adult sizes as early as September or October.

The high growth rates are consistent with ruderal strategy, and, therefore, by this parameter, *H. panicea* and *M. citrina* can be classified as *R*-species.

Hiatella arctica. The 4-year old specimens inhabiting benthic communities in the White Sea, average 25 mm in size (Matveeva, Maximovich, 1977). On the other hand, the modal size of *H. arctica* of the same age inhabiting the fouling communities in the White Sea is only about 8 mm (Khalaman 1993). It should be noted that these data were obtained from *H. arctica* living in the mussel's fouling community, and, therefore, these low growth rates could be explained by an adverse influence of the *Mytilus edulis* (Khalaman, Komendantov, 2007). In any case, *H. arctica*, with its long life span and relatively small definitive sizes, grows at a very slow rate. This characteristic is consistent with the tolerance (*S*) strategy.

Mytilus edulis and *Styela rustica*. Both species have comparably large definitive sizes and prolonged life span. However, the growth characteristics of these two species are different. Under suspended cultivation conditions, which are equivalent to the animals being in the fouling community, the mussels in the White Sea could reach 50 mm in size during the first 3–4 years of life. It was shown that the relationship between the shell length and age in the mussels grown in suspended cultivation is close to a linear one (Sukhotin, Kulakowski, 1992). This is indicative of high growth rates during the first few years of mollusk life, which is especially advantageous for a rapid occupation of exposed surfaces and could be qualified as a characteristic of *R*-species.

By contrast, *S. rustica* grows at a slow rate during the first two years of its life. One-year old ascidians have L_{comb} = 1.7±0.16 mm, two-year old ascidians L_{comb} = 3.5±0.34 mm (Khalaman, in press). This is one of the main reasons, why *S. rustica* does not predominate in the fouling communities on the substrates, exposed for a period of less than 2-3 years (Oshurkov, 1992; Khalaman, 2001a, b). Moderate growth rates and the lack of the need for a rapid full seizure of exposed surface is, in my opinion, a characteristic of C-species. The settlement of this kind develops at a relatively slow rate, but eventually leads to a full occupation of the substrate.

FERTILITY AND THE VOLUME OF LARVAL POOL

Mytilus edulis. This is probably the most fertile species of all the macrofoulers in the White Sea. The fertility of these animals depends on both the habitat (littoral or sublittoral zone) and the body size. The mean individual fecundity of the White Sea mussels inhabiting the littoral zone is 350000 oocytes/specimen, and of those living in the sublittoral zone is 500000 oocytes/specimen (Maximovich, Gerasimova, 1997).

The concentration of mussel larvae in the plankton in certain areas of the White Sea can exceed 1000 specimens/m^3 and reach 18 700 specimens/m^3 (Shilin, Oshurkov, 1985; Maximovich, Vedernikov, 1986; Maximovich, Shilin, 1991, 1993).

Hiatella arctica. There is no information in the literature concerning the fertility of these mollusks. However, during the peak of the spawning period in the White Sea, the larvae of *H. arctica* abound in plankton. The numbers of these larvae are only 2-3 times lower than those of mussel larvae (Flyachinskaya L.P. personal communication). According to other authors, the abundance of *H. arctica* larvae in the Kandalaksha Gulf of the White Sea varies from 50–90 specimens/m^3 to 400 specimens/m^3 (Shilin, Oshurkov, 1985; Maksimovich, Shilin, 1991, 1993). These figures are very high, especially given the smaller definitive sizes of these mollusks compared to those of *M. edulis* and lower abundance of *H. arctica* in the benthic and fouling communities.

Styela rustica and *Molgula citrina.* There is no information in the literature concerning the fertility of both species. According to the author's own observations of *S. rustica* and *M. citrina,* living in aquaria, the number of eggs, laid by a single specimen of *S. rustica*, is low and could be estimated to be several hundred. The fecundity of *M. citrina* is even lower. The maternal specimen releases several dozen of well-developed large larvae.

Halichondria panicea. Despite a large number of studies of this very common species, some of them dedicated to reproduction, there is almost no information on fertility. According to Yu.I. Mukhina (personal communication), 1 cm^3 of basal layer in *H. panicea* contains 90–130 larvae. Consequently, the fecundity of a large-sized colony is approximately 25000–40000 larvae.

The results presented in this section show that the species examined in this paper could be arranged in order of increasing fertility, as follows: *M. citrina* – *S. rustica* – *H. panicea* – *H. arctica* – *M. edulis*. The high fertility is a characteristic of *R*-strategies, and the low fertility is an attribute of *C*-strategists. Therefore, *H. arctica* and *M. edulis* could be classified as *R*-species, whereas *M. citrina* and *S. rustica* - as *C*-species. *H. panicea* occupies an intermediate position in terms of fertility and could be assigned to any of these groups.

The fertility, in my opinion, does not determine the presence or lack of tolerance (*S*) strategy in this species. Therefore, none of the species examined in this paper can be classified by this parameter as *S*-species.

TOLERANCE TO CHANGES IN MAIN ENVIRONMENTAL FACTORS (TEMPERATURE, SALINITY)

Mytilus edulis. Ecological plasticity of mussels are well known (Bayne, 1977). In the White Sea, *M. edulis* lives both in littoral and sublittoral zones, including estuaries, most strongly affected by substantial fluctuations in temperature and salinity.

In winter, the mussels living in the White Sea survive subzero temperatures, and in summer these mollusks living in the littoral zone could withstand heating up to +44°C (Berger et al., 1999). The mussels, acclimated to 25–26‰ salinity, were shown to maintain 100% activity, when salinity is decreased to 14‰, and after one-month acclimation to 10‰ salinity, the activity is maintained, when salinity is dropped to 8‰ (Berger, Lukanin, 1985). If salinity of the surrounding water becomes considerably reduced, the mussels seal off the mantle cavity, by pressing the shell valves together, which allows them to survive adverse conditions for long periods of time. Closed mussels could survive in fresh water for a period of 12 days (Berger, 1980).

Hiatella arctica. The author is unaware of any data concerning the temperature tolerance of this species. In the benthic communities of the White Sea, however, *H. arctica* was detected in the temperature range of –0.9 to 20.3°C and salinity range of 16.4 to 30.1‰ (Naumov, 2006). *H. arctica* is unable to isolate itself from the environment due to the specific shell shape, which has an adverse effect on the ability of this species to withstand fluctuations in salinity. Different authors give more or less similar estimates for the potential salinity tolerance of *H. arctica*, living in the White Sea: 12–37.5‰ (Filippov et al., 2003) and 14–40‰ (Komendantov et al., 2006). It should be noted that the settlement density and biomass of this species in the fouling communities living on artificial substrates, that reside in the surface water layer (0 – 0.5 m), most strongly affected by the annual springtime decrease in salinity, are normally very low (Khalaman, Kulakowski, 1993).

Styela rustica. There is no information concerning the temperature tolerance of this ascidian species. However, judging from its geographical distribution (Redikortsev, 1916; Lützen, 1960; Jewett, Feder, 1981; Kessler, 1985), this is a cold-tolerant species. Abundant benthic settlements of *Styela rustica* normally occur in the White Sea at depths greater than 5 – 10 m (Golikov et al., 1985), i.e., below the summer thermocline zone.

According to the author's own observations, heating to 18°C is lethal for *S. rustica*. During the unusually warm hydrological summer of 2000, the author observed mass mortality of these animals in one of the shallow bays of the White Sea. The ascidians grew on artificial substrates suspended at depths of 1.5 - 3 m. Water temperature at a depth of 3 m was 17–18°C, and the surface water layer was warmed up to 20–22°C.

Salinity tolerance range for *S. rustica* living in the White Sea is unknown. However, the lower tolerance limit of salinity for these ascidians, acclimated to 24‰, varies from 10‰ to 15‰. This is true, provided that the animals were exposed to this salinity and the water temperature of +11°C, at least, for a period of 150 hours. A prolonged stay in the water, with a salinity of 15‰ or higher, has no apparent negative effect on these ascidians (Khalaman, Isakov, 2002).

Even a short (15 min to 0.5 hour) immersion in fresh water is fatal for *S. rustica*, and, at 5‰ salinity, mass mortality of ascidians begins within 30 min. On the other hand, *S. rustica* can survive several hours, if water salinity is 10‰ (Khalaman, Isakov, 2002). In this respect, these ascidians can be classified as marine poikilosmotic animals, unable to isolate themselves from the environment. The White Sea representatives of the Nudibranchia and Pteropoda, for instance, respond to decrease in salinity in a similar way. They die at low salinity (0–8‰) within 10–30 min. At salinity over 10‰, their survival time is greatly increased (Berger, 1986).

The springtime decrease in salinity in the surface water layer is likely to prevent *S. rustica* from reaching above 1 – 1.5 m. A similar limiting role of the periodic decrease in water salinity is also reported for another member of the family Styelidae, *S. plicata.* During the monsoon rains, the salinity in the upper 2 m of the water column in the Sea of Japan drops below 14‰, which prevents ascidians from reaching these depths (Kazihara, 1962).

Molgula citrina. There is no information concerning the temperature tolerance of this species. The salinity tolerance range for adult individuals of

M. citrina, living under normal salinity conditions of the White Sea (24‰) is fairly narrow and is 20–30‰ at +10°C. After a prior acclimation to 20‰ salinity, *M. citrina* can survive a salinity range of 14‰ to 30‰ for 5 days (Komendantov, Khalaman, unpublished data). *M. citrine*, however, is less tolerant to low salinity, than *Styela rustica*.

The specimens of *S. rustica* remain viable, when transferred from the normal salinity of the White Sea (24‰) to the salinity of 15‰ (Khalaman, Isakov, 2002), whereas *Molgula citrina*, acclimated to 24‰, remain unharmed, when transferred into the water of not less then 20‰ salinity.

Halichondria panicea. There is no direct information in the literature on the temperature and salinity tolerance of this species. However, this sponge is quite common in various climatic zones. It was detected in the Arctic and Antarctic, Atlantic, Pacific and Indian Oceans (Althoff et al., 1998). Taking into account these circumstances, *H. panicea* could be regarded as a sufficiently eurybiontic species.

In the Barents Sea, this sponge is found in the littoral zone (Yereskovski, 1994). In the White Sea, however, *H. panicea* is only occasionally found in the intertidal zone and only in its lower level, on the brown algae. According to the author's own observations, in the fouling communities of the White Sea, *H. panicea* keeps off the depths less then 0.5 m.

A relatively high tolerance of *H. panicea* is supported by the unpublished data on the survival range of the larvae, which is 12–35‰ (O.L. Saranchova, personal communication).

By the parameter of response of the organisms examined in this paper to the changes in environmental factors, *M. edulis* could be regarded as eurybiontic species; *S. rustica*, *H. arctica* and *H. panicea* could be classified as moderately eurybiontic, and *M. citrina* – as stenobiontic. Moderate eurybionty could be considered a characteristic of *C*-species, high – of *S*-species, and low – of *R*-species.

EDIFICATORY ABILITIES OF MONOSPECIFIC SETTLEMENTS

Mytilus edulis. The ability of mussels to live in tight aggregations and to form extensive continuous settlements, so called mussel beds, is well known. Large littoral and sublittoral mussel settlements are common in the White Sea, especially in its northwest portion, the Kandalaksha Gulf (Lukanin, 1985).

The mussel settlements are able to substantially modify the surrounding environment both in benthic and fouling communities. *M. edulis* plays a role of a species that generates living conditions for a whole assemblage of the organisms, which are in some way associated with the mussels (Reish, 1964b; Tsuchia, Nishihira, 1986; Ardizzone et al., 1989; Khalaman, 1989, 1998, 2001a; Lintas, Seed, 1994; Seed, 1996; Svane, Setyobudiandi, 1996; .Khaitov et al., 2007 and many others).

Hiatella arctica. The fouling communities dominated by *H. arctica,* are very rare in the upper 3 m of the water column in the White Sea (Khalaman, 2001a, 2005b).

The reason is likely to be the competition with larger animals, such as *Mytilus edulis* and *Styela rustica.* In addition, *H. arctica* has only a very limited ability to form aggregations similar to mussel patches. This animal is sluggish and leads a cryptic lifestyle preferring to settle in sheltered places (Naumov, 2006). Even so, *H. arctica* remains the principal subdominant member of the fouling communities (Khalaman, 1993, 2001a; Maximovich, Morozova, 2000).

The fouling communities dominated by *H. arctica,* are common at the depths over 5 m. Assemblage of *Hiatella arctica + Bryozoa gen. sp.* is typical variant of it. However, *Hiatella* community is replaced in succession by other communities in the White Sea (Oshurkov, 1985).

It is interesting that in the North Atlantic, on new lava grounds of the volcano Jan Mayen, the settlements of *H. arctica,* occuring in the community together with bryozoans, are also very common (Gulliksen et al., 1980). Consequently, although *H. arctica* has the ability to form mass settlements, but this ability is most often "obscured" by other foulers.

The cryptic lifestyle, small definitive sizes, "loose" and single-layered architecture of aggregations are the reasons, why *H. arctica* is likely to have only limited edificatory abilities. According of own observation the diversity and abundance of associated organisms in community dominated by *H. arctica* are low.

Styela rustica. High edificatory properties of the settlements of large solitary ascidians have been described (Monteiro et al., 2002). In the White Sea, *S. rustica,* together with the mussels, is one of the species able to predominate in the long-term communities and to define the appearance of the community as a whole. *S. rustica* is capable of forming tight clusters: patches and beds, typical not only for the fouling communities living on the artificial substrata (Khalaman, 1998; 2001a), but also for the mollusk shells, barnacle and underwater rocks (Yakovis, 2005; 2007). A great number of various organisms live on the aggregations of *S. rustica.* Species diversity in the associated flora and fauna of the fouling assemblages, formed by *S. rustica,* is higher than that in the *Mytilus* communities (Khalaman, 2001a). Thus, it could be presumed that the role of

S. rustica in modifying the surrounding environment and in organizing the community is very high.

Molgula citrina. This ascidian species could form vast conglomerates in the fouling communities of the White Sea, comprising a great number of specimens. If such a conglomerate develops on a flat or slightly curved surface, it forms a sheet of tightly-packed individuals, arranged in a single layer. Tight clusters of *M. citrine* normally lack associated macrofauna. In the long-term fouling communities, *M. citrina* colonize other organisms: hydroids (*Obelia longissima*), bivalves (*Mytilus edulis*), ascidians (*Styela rustica*) and brown algae (*Laminaria saccharina*) and could be qualified as epibiont. The ability of this species to create the living conditions for other organisms is most likely to be very limited. In contrary, *M. citrina* itself can take advantage of the resources (substrate), provided by other organisms.

Halichondria panicea. The sponge colonies are capable of generating the living conditions for a rich associated fauna (Frith, 1976; Westinga, Hoetjes, 1981; Çinar et al., 2002; Abdo, 2007). *H. panicea* living in the White Sea are no exception to this rule.

According to the author's own observations, the most common species, occurring together with this sponge, are the polychetes *Nereis pelagica* and *Harmothoe imbricata* and the brittle star *Ophiopholis aculeata* (Khalaman, 2001a). It should also be noted that the individuals of *N. pelagica* live in the irrigation system of *H. panicea*.

In the fouling assemblages of the White Sea, *H. panicea* is distributed in a highly mosaic manner. The sizes of colonies are normally limited by 2-3 dm^3, and unlike mussels and ascidians, *H. panicea* does not cover wide areas. The reason most likely being that the active growth of *H. panicea* occurs only during the first half of the summer, and, during the fall, the major part of the colony becomes degraded.

However, in the places, occupied by *H. panicea*, all other sedentary organisms become suppressed. The sponge intensively overgrows both mollusks and ascidians, very often covering them completely.

The ability to serve as the edificatory member of the community should be characteristic of *C*-species. Consequently, by this parameter, *Mytilus edulis*, *Styela rustica* and *Halichondria panicea* display competitor strategy. The lack of this ability or low ability should be observed in both *R*- and *S*-species. Both *Molgula citrina* and *Hiatella arctica* belong to this category. By this parameter, they should be classified as having either tolerance or ruderal strategy.

ATTACHMENT STRENGTH

The main resource, for which the organisms are competing in the epibenthic communities, is the available open space (Dayton, 1971; Jackson, 1977; Buss, 1986; Nandakumar, Tanaka, 1993 and many other). Consequently, the strength, with which an animal is attached to the substrate, is one of the main competitive properties. Under otherwise equal conditions, the organism that wins the competition is that, which is more difficult to mechanically move or crowd out from its current place.

Mytilus edulis. Although these animals grow in tight clusters, they have a relatively low ability to attach to the substrate. The adherence of the mussels to the substrate through the byssal threads is not permanent. The animals are constantly renewing their byssus and change their position (Lee et al., 1990), although their ability to move around in tight settlements is limited (Paine, 1974; Okamura, 1986). As individual animals grow in size, their adherence to the substrate is weakened and so the mussel patches could come off under their own weight from the vertically-oriented substrates, especially, under the action of waves (Maximovich et al., 1993).

Hiatella arctica. The strength of attachment by byssal threads is much lower in this mollusk, than that in mussels, but their ability to penetrate into a variety of sheltered places helps them to remain successfully settled in the fouling communities.

Styela rustica. This species is very tightly attached to various surfaces. Even when the animal dies, it remains attached to the vertically oriented substrates until the tissues completely dissolve.

Molgula citrina is very loosely attached to the substrate. It comes off easily from any surface both separately, and in whole groups of tightly fused individuals.

Halichondria panicea is characterized by a tight mechanical connection to the substrate, and is capable of penetrating into all cavities and cracks of the substrate.

To sum up, the species of fouling organisms examined in this paper could be roughly divided into three groups: the species firmly affixed to the substrate (*S. rustica, H. panicea*), the species, only loosed attached to the substrate (*M. citrina*), and those occupying an intermediate position (*M. edulis, H. arctica*). The high degree of adherence to the substrate should be characteristic of *C*-species. The moderate degree is possible for both *R*- and *S*-species. The low degree of adherence is the property of *R*-species.

PUMPING RATE

There is a belief that competition for food (both intra- and interspecific) is not the major driving factor in epibenthic communities (Lesser et al., 1992; Lohse, 2002). The sizes of food particles preferred by the competing species are often quite different, which reduces the tension in these relations (Mook, 1981b; Stuart, Klumpp, 1984). Nevertheless, the struggle for food resources could have a great importance for sedentary animals (Buss, 1981; Zajac et al., 1989; Dalby, Young, 1993; Leblanc et al., 2003). The performance of filter-feeding organisms in terms of food competition is determined by a combination of several factors, such as sedimentation efficiency, the range of the food spectrum, food assimilability, and the ability to intercept water flows. The latter being directly dependent on the pumping rate of the animals. A series of specially designed studies were conducted to compare the fouling organisms examined in this paper, using this parameter (Lezin et al., 2006; Lezin, Khalaman, 2007).

It turned out that this comparison is fairly difficult to make. The reason being that in *M. edulis, H. arctica* and *S. rustica* the relationship between the pumping activity and body weight is described by a linear function, whereas in *H. panicea* and *M. citrina* this function is exponential (Figure 3). It is, however, obvious that *H. arctica* has the lowest activity of all tested species, whereas *M. edulis* and *H. panicea* the highest. *S. rustica* occupies an intermediate position. The maximum performance of *M. citrina* is limited by the small size of this animal, but could probably be estimated as moderate.

The maximum pumping rate should be characteristic of C-species, a moderate one is common in *R*-species, and the lowest values of this parameter are characteristic of *S*-species, because the organisms, that have this latter strategy, should be satisfied even by scarce resources.

To sum up, when the species were classified using this parameter, *M. edulis* and *H. panicea* were found to have the competitive strategy, *S. rustica* and *M. citrine* - ruderal strategy , and *H. arctica* - tolerance strategy.

RESISTANCE TO INVASION OF OTHER FOULING ORGANISMS

This characteristic, unlike those examined earlier, is a complex (not an elementary) property. It is produced by an interaction of certain abilities, including some of those mentioned earlier. Since each of the fouling species has **individual manner to solve this problem, the use of** "resistance to invasion of other organisms" as one of the parameters is a forced formalization of a combined manifestation of various elementary properties.

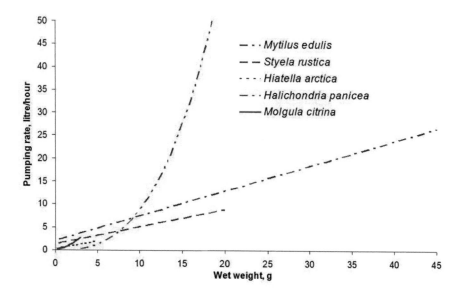

Figure 3. Regression lines for pumping activity vs. wet weight in fouling species tested (by Lezin et al., 2006; Lezin, Khalaman, 2007).

Mytilus edulis. The mussels are thought to have very high competitive ability in epibenthic communities (Paine, 1984). A prolonged existence of mussel settlements on the artificial substrates in the White Sea (Oshurkov, 1985, 1992) indicates that these animals could efficiently prevent the invasion of other sedentary organisms. According to the author's own observations, in the fouling communities of the White Sea, the mussels aged less than 4–5 years almost completely lack epibionts attached to their shells. High filtration activity and tight clustering of these animals (Lezin, 2007) facilitate the olygomixity in the *Mytilus* communities. Excretory-secretory products of these mollusks could also have an adverse effect on other foulers. It was found, for example, that the chemical cues released by mussels into the environment substantially impede the settlement and metamorphosis of the *S. rustica* larvae (Khalaman et al., 2008).

Hiatella arctica. This mollusk leads a cryptic life style. For this reason, *H. arctica* apparently does not need any pronounced ability to put obstacles in the way of other sedentary organisms. Dense and vast settlements of this mollusk, equivalent to those, created by the mussels or *Styela rustica,* are a fairly rare and probably specific phenomenon, determined by the lack of other contenders for the substrate. Moreover, excretory-secretory products of *H. arctica* have positive effect on the larval settlement of the mussels and especially the ascidian *S. rustica*. (Khalaman et al., 2008, 2009).

Styela rustica. Even though it is known that some species of ascidians have developed chemical protection against epibionts and competitors (Stoecker, 1978, 1980; Wahl, Banaigs, 1991), no adverse effect of excretory-secretory products of *S. rustica* on the settlement of mussel larvae was observed (Khalaman et al., 2009). On the other hand, the juveniles of mussels settled down on the settlement of *S. rustica*, are subsequently eliminated, and the survived specimens are characterized by low growth rates (Maximovich, Morozova, 2000; Khalaman, 2005a). These facts could explain the ability of *S. rustica* to successfully resist the invasion of competitors. The specimens of *S. rustica,* that have small or intermediate sizes, lack epibiotic encrustation, but larger ascidians are almost always covered by an epibiotic assemblage, composed predominantly of various filamentous algae. *S. rustica* is often colonized by the hydroid *Obelia longissima*, but rarely by bryozoans, conspecific ascidians or by *Molgula sp*. Mollusks and barnacles are absent. Just like the mussel settlements, the settlements of *S. rustica* in the fouling communities on artificial substrates remain very stable in time (Khalaman, 2005a). All these facts corroborate the ability of these ascidians to successfully resist the invasions of other sedentary organisms.

Molgula citrina. These ascidians are capable of producing the whole strata composed of many individuals almost completely fused together. These formations could probably mechanically prevent the larvae of the foulers from reaching the substrate. The ascidians themselves are usually completely free of epibionts. This could be explained by small life span of this species; however, the excretory-secretory products of *M. citrina* could also impede to some extent the metamorphosis of the *S. rustica* larvae (Khalaman et al., 2008).

At the same time, *M. citrina* is hardly able to inhibit the invasion of (especially, long-lived) organisms. This is explained by the short life span of these animals, and the low durability and short longevity of their settlements. Consequently, the ability of *M. citrina* to resist the invasion of other sedentary organisms could be estimated as moderate.

Halichondria panicea. This sponge takes an active part in the formation of epibiosis of sessile organisms (Yereskovski, Semenova, 1988), but the sponge itself is completely devoid of epibiotic encrustation. *H. panicea* has developed efficient mechanisms of protection against epibionts through periodic sloughing of the surface tissue layer (Barthel, Wolfrath, 1989) and release of toxic secondary metabolites into the environment (Althoff et al., 1998; Kobayashi, Kitagawa, 1998; Dobretsov et al., 2005). Excretory-secretory substances of the White Sea

H. panicea stimulate the larva settlement of the ascidian *Styela rustica* and bivalve *Mytilus edulis*, however, they also inhibit their metamorphosis and induce tissue necrosis. Most larvae die as a result of exposure to these chemicals (Khalaman et al., 2008, 2009). It is interesting to note that a similar effect on the ascidian larvae has the cytotoxic alkaloid extracted from another sponge, *Haliclona* sp. (Green et al., 2002). To sum up, *M. edulis, S. rustica* and *H. panicea* could be regarded as organisms that have a very strong ability to resist the invasion of the competitors. *M. citrina* possesses these attributes to a moderate, and *H. arctica* to a very slight extent. The ability to resist the invasion of other foulers is a characteristic of *C*-species, whereas the absence of such could signify the presence in a species of either tolerance or ruderal strategy.

SPECIES POSITION IN RAMENSKII-GRIME COORDINATE SYSTEM

It should be noted that an attempt to define species position in the *C-S-R* system, made in this paper, is different from the triangular ordination method, proposed by Grime (1979). According to the Grime's model, the continuum of possible strategies is limited by the sides of the equilateral triangle, whose apexes are occupied by absolute *C-*, *S-* и *R-* strategies. The base and the right side of the triangle represent the scales: the potential rate of dry matter production (R_{max}) and morphology index (M), respectively; the latter describing the structural characteristics of the plants.

It was impossible to extrapolate this scheme to marine sedentary animals. One of the main reasons lies in the inapplicability of the morphology index, devised for plants, to animal morphology. Moreover, morphological approach to the animals with different body plans (bivalves, sponges and ascidians), in my opinion, is too complicated and shows little promise. Therefore, the evaluation of the fouling animals was conducted by using the parameters, that reflect not only morphological, but also some functional characteristics of these species.

The estimates for the various parameters, received by each species, were summarized in a tabular form and were used as coordinates to plot the species positions in *C-S-R* space, where each of the strategies was depicted as a vector in 3-dimensional Cartesian coordinate system (Figure 4).

Table. Characteristics of fouling organisms

Species	M. edulis	S. rustica	H. arctica	M. citrina	H. panicea
Definitive size	+, C	+, C	-, R/S	-, R/S	+, C
Life span	+, C/S	+, C/S	+, C/S	-, R	-, R
Growth rate	+, R	+/-, C	-, S	+, R	+, R
Fertility	+, R	-, C	+, R	-, C	+/-, C/R
Tolerance to environmental factors	+, S	-/+, C	-/+ C	-, R	-/+, C
Edificatory ability of settlement	+, C	+, C	-, R/S	-, R/S	+, C
Strength of attachment	+/-, R/S	+, C	+/-, R/S	-, R	+, C
Pumping rate	+, C	+/-, R	-, S	+/-, R	+, C
Resistance to invasion of other fouling organisms	+, C	+, C	-, R/S	+/- C	+, C
Sum	5C, 3S, 3R	8C, 1S, 1R	2C, 7S, 5R	1C, 3S, 7R	7C, 0S, 3R

Note: + – high level, - – low level, +/- – moderate level; C – competitive strategy; S – tolerance strategy; R – ruderal strategy.

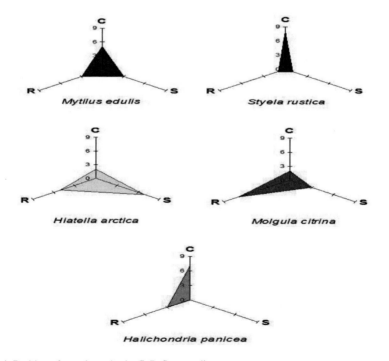

Figure 4. Position of tested species in C, R, S – coordinate system.

Mytilus edulis and *Styela rustica*. As seen from the table and figures (Figure 4) these sedentary organisms display the competitive strategy. They, however, have substantial differences. The mussel, apart from the traits, typical for the *C*-species, possesses many attributes of *S*- and *R*-species (Figure 4). Such species should be ecologically flexible. Indeed, this mollusk has a competitive advantage in habitats with periodic disturbances, caused by both abiotic and biotic factors, under relatively unstable conditions (temperature or salinity fluctuations, intensive water dynamics). These conditions determine the development of mussel fouling communities that could exist for an indefinitely long time (Oshurkov, 2000; Khalaman, 2005a).

The ascidian *S. rustica* has very pronounced characteristics of the *C*-species. The elements of the tolerance and ruderal strategies are present in this species only to a slight degree (Figure 4). For this reason, high competitive ability of *S. rustica* is manifested in a sufficiently narrow range of relatively stable abiotic conditions. This conclusion that follows from the analysis of the diagram is fully consistent with the actual role of this animal in the fouling communities of the White Sea. *S. rustica* gains an advantage at greater depths, than does the mussel, i.e., under more stable abiotic conditions (Oshurkov, 1992, 2000; Khalaman 2001b, 2005a).

The presence of two *C*-species (*Mytilus edulis* and *Styela rustica*) condemns long-term fouling communities in the White Sea within a certain range of abiotic conditions to periodical change of the dominant species: ascidia replaces mussel and vice versa (Khalaman, 2005a). In the community dominated by any of these species, the other foulers, characterized by more pronounced features of *S*- or *R*-species, are inevitably forced into a position of subdominant members.

Hiatella arctica. As seen from the diagram (Figure 4), *H. arctica* has the tolerance strategy. The actual position of this species in the structure of the fouling communities is perfectly consistent with these characteristics. This species is a permanent and abundant component of such communities, occupying there the position of subdominant member. The fouling communities are only very rarely dominated by *H. arctica* and only for a relatively short period (Kulakowski et al., 1993; Oshurkov, 2000; Khalaman, 2001a).

The evidence for the presence of tolerance strategy in *H. arctica* was obtained by comparing physiological and biocenotic (community success) optima (Khalaman, 2005b), which, according to some theoretical considerations, should not coincide in the species that accept this strategy (Vasilevich, 1987). The life strategy of *H. arctica* can be corrected and defined as stress- and competitor-tolerant strategy.

Molgula citrina. As seen from the table and diagrams (Figure 4), this ascidian species has the most pronounced properties of *R*-species, although low fertility, release of well-developed larvae and a certain degree of allelopathy are inconsistent with ruderal strategy. At the same time, the position of this ascidian in the fouling communities is quite consistent with the role of *R*-species. *M. citrina* is a typical member of early stages in community development (short-lived organism phase), and in the long-term fouling assemblages it could appear only as an epibiont and species possessing relatively low rang of ability (Khalaman, 2001a).

Halichondria panicea. This sponge has attributes of both *C*- and *R*-species (Figure 4). Grime (1979) admits the existence of mixed, secondary strategies, including the competitive ruderal strategy (*C-R*). On one hand, *H. panicea* is a highly competitive species; but, on the other hand, it has a seasonal development, and therefore, is unable to hold the substrate for very long. As noted earlier, in the places, occupied by *H. panicea*, this sponge suppresses all other sedentary species. It should be noted, however, that the colonies of *H. panicea* in the fouling assemblages are distributed in a mosaic manner and do not extend as continuous growths over large areas. Thus, *H. panicea* fails to maintain the leading position in the fouling communities of the White Sea, which is consistent with the *C-R* strategy that was determined for this sponge.

It should be emphasized that the strategies of the species listed above were determined based on biological and autoecological characteristics of these organisms. I tried to avoid those parameters that show in some way the species position in the development and in the structure of the fouling communities. Unfortunately, this principle cannot always be strictly followed. A certain exception to this principle is the parameters "edificatory ability of settlement" and "resistance to invasions of other fouling organisms". The choice of these parameters was dictated by their importance for the assessment of life strategies of fouling organisms.

The species characteristics are entirely consistent with the actual position of these organisms in the structure of fouling communities and with their role in the succession. It can, therefore, be concluded that the Ramenskii-Grime system, devised for terrestrial plants, is fairly applicable to the marine epibenthic communities. The results presented in this paper are very important, because they demonstrate the applicability of a single concept to entirely different ecosystems. In turns, this shows a certain commonality of ecological patterns, regardless of the systematic position of the organisms in question and the environment, in which they live, whether it be land or water.

The approach proposed in this paper is based on the ranked assessment method and is not a strict mathematical model. The number and matter of the parameters selected for analysis may be subject to argument. I have selected nine parameters.

The principles used to assign life strategies to certain properties need to be further verified. This aspect of the present study is also open to criticism and further improvement. In this paper I tried to outline the approach for solving of such fundamental problem as explanation of structure and development of epibenthic communities.

Despite the apparent imperfection mentioned above, the life strategy assessment method for sedentary organisms, even in its current form as presented in this paper, is perfectly suitable for analyzing the structure and dynamics of the fouling communities. Knowing the strategy of a species, its biotic environment and abiotic conditions, it is easy to predict, in a sufficiently reliable manner, the further course of development and composition of the developing community.

The approach proposed in this paper allows us to extrapolate the Ramenskii-Grime system to other types of marine communities. This system could, for instance, be applied to the communities living on soft bottoms. In this case, the parameters chosen to evaluate life strategies should be modified. The parameters selected should be essential for the main ecomorphs that form the communities under evaluation.

One of the factors limiting wider application of the Ramenskii-Grime system is the necessity to have detailed information about the biology of the organisms used for analysis.

Unfortunately, there is only a relatively small number of model species or species of practical value among the marine invertebrates, sufficiently well studied to be used for analysis. Among the benthic animals living in the White Sea, there are only a few such species: *Asterias rubens, Mytilus edulis*, the mollusks of the genus *Littorina, Macoma balthica, Mya arenaria, Nereis virens* and some others.

For many organisms, there is a lack of even very basic data: such as the information of food spectrum, life span, etc. (see Naumov, 2006) A series of specially designed studies were conducted to evaluate the five fouling species, according to the parameters selected for analysis (Khalaman, Isakov, 2002; Filippov et al., 2003; Komendantov et al., 2006; Lezin et al., 2006; Khalaman, Komendantov, 2007; Lezin, Khalaman, 2007; Khalaman et al., 2008, 2009, and others). Unfortunately, in some cases, the species had to be evaluated, using only indirect information.

Although the conventional research methods, based on the field observations and sampling, will certainly remain both necessary and essential to study the development and structure of the benthic communities, the techniques, similar to the Ramenskii-Grime system, that help explain and predict the processes, taking place in the communities, will be the next logical step in addressing these problems.

ACKNOWLEDGMENTS

I wish to sincerely thank all my colleagues and coauthors, without whose dedicated daily work and assistance this study would have been impossible. The most part of study was carried out with the framework of projects (03-04-49701, 06-04-48789, 07-04-00854, 10-04-00310) granted by Russian Fundation of Basic Research.

REFERENCES

Abdo D.A. Endofauna differences between two temperate marine sponges (Demospongiae ; Haploscleridae; Chalinidae) from Southwest Australia. *Mar. Biol.*, 2007, 152, 845-854.

Althoff K; Schutt C; Steffen R; Batel R; Muller WEG. Evidence for a symbiosis between bacteria of the genus *Rhodobacter* and the marine sponge *Halichondria panicea*: harbor also for putatively toxic bacteria? *Mar. Biol.*, 1998, 130, 529-536.

Anger K. Development of a subtidal epifaunal community at the island of Helgoland. *Helgol. Wiss. Meeresunters.* 1978, 31, 457-470.

Ardizzone GD; Gravina MF; Belluscio A. Temporal development of epibenthic communities on artificial reefs in the central Mediterranean Sea. *Bull. Mar. Sci.* 1989, 44, 592-608.

Babkov AI; Golikov AN. Hydrobiocomplexes of the White Sea. Leningrad. *Zoological Institute Academy of Sciences USSR.* 1984. (in Russian).

Bailey-Brock JH. Fouling community development on an artificial reef in Hawaiian waters. *Bull. Mar. Sci.* 1989, 44, 580-591.

Barkai A; Branch GM. The influence of predation and substratal complexity on recruitment to settlement plates: a test of the theory of alternative states. *J. Exp. Mar. Biol. Ecol.* 1988, 124, 215-237.

Barthel D. On the ecophysiology of the sponge *Halichondria panicea* in Kel Bight. I. Substrate specificity, growth and reproduction. *Mar. Ecol. Progr. Ser.* 1986, 32, 291-298.

Barthel D; Wolfrath R. Tissue sloughing in the sponge *Halichondria panicea* a fouling organism prevents being fouled. *Oecologia.* 1989, 78, 357-360.

Bayne BL. (ed.) Marine mussels: there ecology and physiology. London. 1977.

Berger VYa. Euryhalinity of marine mollusks: Ecological, morphofunctional and evolutionary aspects. Leningrad. *Zoological Institute AS USSR.* 1980. (in Russian).

Berger VYa. Adaptations of Marine Molluscs to Environmental Salinity Changes. Exploration of the fauna of the seas. *Zoological Institute AS USSR.* 1986, 32(40). (in Russian).

Berger VYa.; Sokolova IM; Sukhotin AA. Morphological and behavioral aspects of adaptations to the summer heating in the inhabitants of the intertidal zone of the White Sea. In: Annual reports of Zoological Institute RAS. St. Petersburg. *Zoological Institute RAS.* 1999, 13. (in Russian).

Berger VYa; Lukanin VV. The adaptive reactions of White Sea blue mussel on environmental salinity changes. In: Scarlato O.A. ed. Investigations of the mussels of the White Sea. Leningrad. *Zoological Institute AS USSR.* 1985, 4-21. (in Russian).

Berger V; Dahle S; Galaktionov K; Kosobokova K; Naumov A; Rat'kova T; Savinov V; Savinova T. White Sea. Ecology and Environment. St. Petersburg – Tromsø. Derzavets Publisher. 2001.

Begon M; Harper JL; Townsend CR. Ecology: individuals, populations and communities. 2nd edn. Oxford. Blackwell Scientific Publications. 1990.

Bowden DA; Clarke A; Peck LS; Barnes DKA. Antarctic sessile marine benthos: colonization and growth on artificial substrata over three years. *Mar. Ecol. Prog. Ser.* 2006, 316, 1-16.

Brayko VD. Fouling in the Black Sea. Kiev. Naukova Dumka. 1985. (in Russian)

Brayko VD; Dolgopol'skaja VD. The principal characteristics of the formation of fouling cenosis. *Hidrobiologicheski zhurnal.* 1974, 10, N1, 11–18. (in Russian).

Brown KM; Swearingen DC. Effect of seasonality, length of immersion, locality and predation on an intertidal fouling assemblage in the Northern Gulf of Mexico. *J. Exp. Mar. Biol. Ecol.* 1998, 225, 107-121.

Breitburg DI. Development of a subtidal epibenthic community: factors affecting species composition and the mechanisms of succession. *Oecologia.* 1985, 65, 173-184.

Buss LW. Group living, competition, and the evolution of cooperation in a sessile invertebrate. *Science.* 1981, 213, 1012-1014.

Buss LW. Competition and community organization on hard surfaces in the sea. In: Diamond J; Case T.J. ed. Community ecology. New York. Harper and Row. 1986, 517-536.

Butler AJ; Connolly RM. Development and long tern dynamics of a fouling assemblage of sessile marine invertebrates. Biofouling. 1996, 9, P. 187-209.

Butler AJ; Connolly RM. Assemblages of sessile marine invertebrates still changing after all these years? *Mar. Ecol. Progr. Ser.* 1999,182, 109-118.

Campbell BD; Grime JP. An experimental test of plant strategy theory. *Ecology.* 1992, 73, 15-29.

Çinar ME; Katağan T; Ergen Z; Sezgin M. Zoobenthos-inhabiting *Sarcotragus muscarum* (Porifera: Demospongiae) from the Aegean Sea. *Hydrobiologia.* 2002, 482, 107-117.

Clements FE. Plant succession. An analysis of the development of vegetation. Washington D.C. *Carnegie Inst. Wash. Publ.* 1916. N 242.

Connell JH; Slatyer RO. Mechanisms of succession in natural communities and their role in community stability and organization. *Am. Nat.* 1977, 11, 1119-1144.

Dalby JE; Young CM. Variable effects of ascidian competitors on oysters in a Florida epifaunal community. *J. Exp. Mar. Biol. Ecol.* 1993, 167, 47-57.

Dayton P.K. Competition, disturbance and community organization : The position and subsequent utilization of space in a rocky intertidal community. *Ecol. Monogr.* 1971, 41, 351-389.

Dean TA. Structural aspects of sessile invertebrates as organizing forces in an estuarine fouling community. *J. Exp. Mar. Biol. Ecol.* 1981, 53, 163-180.

Dean TA; Hurd LE. Development in an estuarine fouling community the influence of early colonists on later arrivals. *Oecologia.* 1980, 46, 295-301.

Dobretsov S; Dahms H-U; Qian P-Y. Antibacterial and anti-diatom activity of Hong Kong sponges. *Aquatic Microbial Ecology.* 2005, 38, 191-201.

Filippov AA; Komendantov A.Yu; Khalaman VV. Salinity tolerance of the White sea mollusk Hiatella arctica L. (Bivalvia, Heterodonta). *Zoologicheski zhurnal.* 2003, Vol. 82, 913-918. (in Russian).

Fitzhardinge RC; Bailey-Brock JH. Colonization of artificial reef materials by corals and other sessile organisms. *Bull. Mar. Science.* 1989, 44, 567-579.

Frame DW; McCann JA; Growth of *Molgula complanata* Alder and Hancock, 1870 attached to test panels in the Cape Cod Canal. *Chesapeake Science.* 1971, 12, 62-66.

Frith DW. Animals associated with sponges at North Hagling, Hampshire. *Zool. J. Linn. Soc.* 1976, 58, 353-362.

Glasby TM. Development of sessile marine assemblages on fixed versus moving substrata. *Mar. Ecol. Progr. Ser.* 2001, 215, 37-47.

Glasser JW. On the causes of temporal change in communities modification of the biotic environment. *Am. Nat.* 1982, 119, 375-390.

Golikov AN; Skarlato OA; Gal'tsova VV; Menshutkina TV. Ecosystems of the Chupa Inlet of the White Sea and their seasonal dynamics. Exploration of the fauna of the seas. *Zoological Institute AS USSR*. 1985, 31(39), 5-83. (in Russian).

Golikov A.N., Sirenko B.I., Gal'tsova V.V., Golikov A.A., Novikov O.K., Petryashov V.V., Potin V.V., Fedyakov V.V., Vladimirov M.V. Ecosystems of the south-eastern part of the Kandalaksha Bay (White Sea) in the Sonostrov area. Exploration of the fauna of the seas. *Zoological Institute AS USSR*. 1988, 40(48), 4-135. (in Russian).

Green KM; Russell BD; Clark RJ; Jones MK; Garson MJ; Skilleter GA; Degnan BM. A sponge allelochemical induces ascidian settlement but inhibits metamorphosis. *Mar. Biol.* 2002, 140, 355-363.

Grime JP. Evidence for the existence of three primary strategies in plants and its relevance to ecological and evolutionary theory. *Am. Nat.* 1974, 111, 1169-1194.

Grime JP. Plant strategies and vegetation processes. New York. John Willey and Sons. 1979.

Gorin AN. The types of fouling on the floating navigational fence in the north-western part of Sea of Japan. In: Ecology of Fouling in North-Western Part of the Pacific Ocean. Vladivostok. Far East Sci. *Centre Acad Sci USSR*. 1980, 26-31. (in Rusian).

Greene CH; Schoener A. Succession on marine hard substrata: a fixed lottery. *Oecologia*. 1982, 55, 289-297.

Gulliksen B; Haug T; Sandness OK. Benthic macrofauna on new and old lavagrounds at Jan Mayen. *Sarsia*. 1980, 65, 137-148.

Haderlie E. Marine fouling and boring organisms in Monterey Harbor II. Second year of investigation. *Veliger*. 1969, 12, 182-192.

Henschel JR; Cook PA; Branch GM. The colonization of artificial substrata by marine sessile organisms in False Bay. 1. Community development. *S. Afr. J. Mar. Sci.* 1990, N 9, 289-297.

Hirata T. Succession of sessile organisms on experimental plates immersed in Nabera Bay, Izu Peninsula, Japan. II. Succession of invertebrates. *Mar. Ecol. Prog. Ser.* 1987, 38, 25-35.

Hirata T. Succession of sessile organisms on experimental plates immersed in Nabera Bay, Izu Peninsula, Japan. V. An integrated consideration on the defenition and prediction of succession. *Ecological Research*. 1992, 7, 31-42.

Ivanova LV. The life cycle of the Barents Sea sponge *Halichondria panicea* (Pallas). In: Morphogenetic processes in sponges. Leningrad. *LSU*. 1981, 59-73. (in Russian).

Jackson JBC. Habitat area, colonization and development of epibenthic community structure. In: Koegan BF; Ceidigh PO; Boaden PJS eds. Biology of benthic organisms. Oxford. Pergamon Press. 1977, 349-358.

Jewett SC; Feder HM. Epifaunal invertebrates of the continental shelf of the eastern Bering and Chukchi Seas. In: Hood DW; Calder J.A. eds. The Eastern Bering Sea shelf. Oceanography and Resource. Vol. 2. Seattle. Office of Marine Pollution Assessment, NOAA, University of Washington Press. 1981, 1131-1153.

Kazihara T. Influences of the rainfall upon the living of *Styela plicata* (Lesueur). *Bull. Japan. Soc. Sci. Fish.* 1962, 28, 565-569.

Kay AM; Butler AJ. "Stability" of the fouling communities on the pilings of two piers in South Australia. *Oecologia.* 1983, 56, 70-78.

Kawahara T; Iwaki T; Hibino K; Sugimura Y. Fouling communities in Yokkaichi Harbor. *Amakusa Mar. Biol. Lab.* 1979, 5, 19-30.

Kessler DW. Alaska's saltwater fishes and other sea life: a field guid. Anchorage, Alaska. Alaska Northwest Pub. 1985.

Khaitov VM; Artemieva AV; Gornyh AE; Zhizhina OG; Yakovis EL. The role of mussel patches in the community-structuring processes on muddy-sand flats. I. Composition of the patch-associated community in the littoral zone of the White Sea. Vestnik SPbGU. 2007, Ser. 3, Vyp. 4, 3-12. (in Russian).

Khalaman VV. Developmental patterns of the fouling communities growing on artificial substrates and blue mussel (*Mytilus edulis* L.) aquaculture industry in the White Sea. PhD thesis. S. Petersburg. *Zoological institute RAS.* 1993.

Khalaman VV. Studies of succession of fouling communities in the White Sea using the information index of species diversity. *Proceedings of Zoological Institute AS USSR.* 1989, 203, 34–45 (in Russian).

Khalaman VV. Correlations of spatial distribution of organisms in fouling communities of the White Sea. *Journal of General Biology.* 1998, 59. N 1, 58-73 (in Russian).

Khalaman VV. Fouling communities of mussel aquaculture installations in the White Sea. *Russian Journal of Marine Biology.* 2001a, 27, 227-237.

Khalaman VV. Succession of fouling communities on an artificial substrates of a mussel culture in the White Sea. *Russian Journal of Marine Biology.* 2001b, 27, 345-352.

Khalaman VV. Long-term changes in shallow-water fouling communities of the White Sea. *Russian Journal of Marine Biology.* 2005a, 31, 344-351.

Khalaman VV. Testing the hypothesis of tolerance strategies in *Hiatella arctica* L. (MOLLUSCA: BIVALVIA). *Helgoland Mar. Res.* 2005b, 59, N 3, 187-195.

Khalaman VV. Life span and growth rate of *Styela rustica* L. (Ascidiae, Chordata) inhabiting in the White Sea. *Zoologicheski zhurnal*, in press. (in Russian).

Khalaman VV; Isakov AV. The survival of the solitary ascidian *Styela rustica* in the water of low salinity. *Vestnik SPbGU*. 2002, Ser. 3, Vyp. 4, 91-95. (in Russian).

Khalaman VV; Komendantov AYu. Mutual effects of several fouling organisms of the White Sea (*Mytilus edulis*, *Styela rustica* and *Hiatella arctica*) on their growth rate and survival. *Russian Journal of Marine Biology*. 2007, 33, 139-144.

Khalaman VV; Kulakowski EE. The formation of the macrofouling community on the artificial sudstrata in the mussel mariculture conditions in the White Sea. *Proceedings of the Zoological Institute RAS*. 1993, 253, 83-100. (in Russian).

Khalaman VV; Belyaeva DV; Flyachinskaya LP. Impact of excretory-secretory products of some fouling organisms on settling and metamorphosis of the larvae of *Styela rustica* (Ascidiae). *Russian Journal of Marine Biology*. 2008, 34, 170-173.

Khalaman VV; Flyachinskaya LP; Lezin PA. Impact of excretory-secretory products of some fouling organisms on settling of mussel's larvae (*Mytilus edulis* L., Bivalvia, Mollusca). *Invertebrate Zoology*, 2009, 6, N1, 65-72 (in Russian).

Kobayashi M; Kitagawa I. Likely microbial participation in the production of bioactive marine sponge chemical constituents. In: Watanabe Y; Fusetani N. eds. Sponge sciences: multidisciplinary perspectives. Tokyo. Springer-Verlag. 1998, 379-389.

Komendantov AYu; Bakhmet IN; Smurov AO; Khalaman VV. The effect of salinity changes on the heart rate and salinity tolerance in *Hiatella arctica* L. (Bivalvia, Heterodonta), Vestnik SPbGU. Ser. 3, 2006, N 4, 17-24. (in Russian).

Kulakowski EE; Sukhotin AA; Khalaman VV. The formation of cultured mussel populations in different points of Chupa Inlet (Kandalaksha Bay). *Proceedings of the Zoological Institute RAS*. 1993, 253, 24-41. (in Russian).

Leblanc AR; Landry T; Miron G. Fouling organisms of the blue mussel *Mytilus edulis*: Their effect on nutrient uptake and release. J. *Shellfish Research*. 2003, 22, 633-638.

Lee CY; Lim ShSL; Owen MD. The rate and strength of byssal reattachment by blue mussels (*Mytilus edulis* L.). *Can. J. Zool.* 1990, 68, 2005-2009.

Lezin PA. The spatial structure of the White Sea mussel (Mytilus edulis) aggregations. Zoologicheski zhurnal. 2007, 86, 163-166. (in Russian).

Lezin PA; Agat'eva NA; Khalaman VV. A comparative study of the pumping activity of some fouling animals from the White Sea. *Russian Journal of Marine Biology*. 2006, 32, 245-249.

Lezin PA; Khalaman VV. Characteristics of filtration activity of the sestonophages in the fouling communities of the White Sea. In: Ecological investigations of the White Sea organisms. Proceedings of 2nd international conference. Saint-Petersburg. *Zoological Institute RAS*. 2007, 67-69. (in Russian)

Lohse DP. Relative Strengths of Competition for Space and Food in a Sessile Filter Feeder. *Biol. Bull.* 2002, 203, 173-180.

Lintas C; Seed R. Spatial variation in the fauna associated with *Mytilus edulis* on a wave-exposed rocky shore. *J. Moll. Stud.* 1994, 60, 165-174.

Luckens PA. Settlement and succession on rocky shores at Auckland North Island, New Zealand. New Zealand Oceanogr. *Inst. Memoir.* 1976, 70.

Lukanin VV. Spatial distribution of mussel *Mytilus edulis* in the White Sea. In: Scarlato O.A. ed. Investigation of the mussel of the White Sea. Leningrad. *Zoological Institute AS USSR*. 1985, 45-58. (in Russian).

Lützen J. The reproductive cycle and larval anatomy of the ascidian *Styela rustica* (L.). Vidnsk. Medd. fra Dansk naturh. *Foren.* 1960, 123, 227-235.

Maximovich NV; Gerasimova NV. On the fecundity of the bivalves in the White Sea. *Vestnik SPbGU*. 1997, Ser. 3, Vyp. 21, 30-37. (in Russian).

Maximovich NV; Minichev YuS; Kulakowski EE; Sukhotin AA; Chemodanov AV. The dynamics of structural and functional characteristics of cultured mussel settlements in the White Sea. *Proceedings of the Zoological Institute RAS*. 1993, 253, 61-82. (in Russian).

Maximovich NV; Sukhotin AA; Minichev YuS. Long-term dynamics of blue mussel (*Mytilus edulis* L.) culture settlements (the White Sea). *Aquaculture.* 1996, 147, 191-204.

Maximovich NV; Morozova MV. Structural peculiarities of fouling communities on substrata of mussel mariculture (White Sea). *Proceedings of biological institute of SPbGU*. 2000, 46, 124-143. (in Russian).

Maximovich NV; Shilin MB. Spatial distribution of mollusk's larvaton in Chupa Inlet (the White Sea). *Proceedings of Zoological Institute RAS*. 1991, 233, 44-57. (in Rusian).

Maximovich NV; Shilin MB. Larvae of the bivalves in the plankton of the Chupa Inlet (White Sea). Exploration of the fauna of the seas. *Zoological Institute RAS*. 1993, 45 (53), Part II, 131-137 (in Russian).

Maximovich NV; Vedernikov VM. Ecology of the larvae of *Mytilus edulis* L. in the Chupa Inlet (White Sea). In: Scarlato O.A. ed.The ecological Investigation of bottom organisms of the White Sea. Leningrad. *Zoological Institute AS USSR.* 1986, 30-35. (in Russian).

Millard N. Observations and experiments on fouling organisms in Table Bay Harbor, South Africa. *Trans. Roy. Soc. South Africa.* 1951, 33, 415-446.

Mirkin BM; Rosenberg GS. Dictionary of modern Phytocenology with Definitions. Moscow. *Nauka.* 1983. (in Russian).

McDougall KD. Sessile marine invertebrates of Beaufort, North Carolina. A study of settlement, growth, and seasonal fluctuations among pile-dwelling organisms. *Ecol. Monogr.* 1943, 13, 321-374.

Monteiro SM; Chapman MG; Underwood AJ. Patches of the ascidian *Pyura stolonifera* (Heller, 1878): structure of habitat and associated intertidal assemblages. *J. Exp. Mar. Biol. Ecol.* 2002, 270, 171-189.

Mook DH. Effects of disturbance and initial settlement on fouling community structure. *Ecology.* 1981a, 62, 522-526.

Mook DH. Removal of suspended particles by fouling communities. *Mar. Ecol. Progr. Ser.* 1981b, 5, 279-281.

Myagkov GM. Composition and distribution of the fauna in the biocoenosis of Laminaria saccharina in the Chupa Inlet (White Sea). *Gidrobiologicheski zhurnal.* 1975, 11, N5, 42-48. (in Russian).

Myers PhE. Space versus other limiting resources for a colonial tunicate, *Botrylloides leachii* (Savigny), on fouling plates. *J. Exp. Mar. Biol. Ecol.* 1990, 141, 47-52.

Nandakumar K. Importance of timing of panel exposure on the competitive outcome and succession of sessile organisms. *Mar. Ecol. Progr. Ser.* 1996, 131, 191-203.

Nandakumar K; Tanaka M. Interspecific competition among fouling organisms. A review. *Publ. Amakusa Mar. Biol. Lab.* 1993, 12, P. 13-35.

Naumov AD. Clams of the White Sea. Ecological and faunistic analysis. Exploration of fauna of the seas. *Zoological Institute RAS.* 2006, 59(67). (in Russian).

Odum EP. Basic Ecology. Vol. 2. Philadelphia. Saunders College Publishing. 1983.

Okamura B. Formation and disruption of aggregations of *Mytilus edulis* in the fouling community of San Francisco Bay, California. *Mar. Ecol. Prog. Ser.* 1986, 30, 275-282.

Olifan VI. Subphylum Tunicata. Class Ascidiacea In: Gaevskaya N.S. ed. Indentification key to fauna and flora of the northern seas of the USSR. Moscow. *Sovetskaya nauka*. 1948, 496-512. (in Russian).

Oshurkov VV. Dynamics and structure of some fouling and benthic communities in the White Sea In: Scarlata O.A. ed. Ecology of fouling in the White Sea. Leningrad. *Zoological Institute AS USSR*. 1985, 44-59. (in Russian).

Oshurkov VV. Succession and climax in some fouling communities. *Biofouling*. 1992, 6, 1-12.

Oshurkov VV. Succession and dynamic of epibenthic communities from the boreal upper subtidal zone. Vladivostok. *Dalnauka*. 2000. (in Russian).

Oshurkov VV; Ivanjushina EA. Epibenthic community succession on the Alaid volcano lavas (North Kuril Islands). *Asian Marine Biology*. 1992, 9, 7-21.

Osman RW. The establishment and development of a marine epifaunal community. *Ecol. Monogr*. 1977, 47, 37-63.

Osman RW; Whitlatch RB. The control of the development of a marine benthic community by predation on recruits. *J. Exp. Mar. Biol. Ecol*. 2004, 311, 117-145.

Paine RT. Intertidal community structure. Experimental studies on the relationship between a dominant competitor and its principal predator. *Oecologia*. 1974, 15, 93-120.

Paine RT. Ecological determism in the competition for space. *Ecology*. 1984, 65, 1339-1348.

Petersen G. H. Life cycles and population dynamics of marine benthic bivalves from the Disco Bugt area of West Greenland. // Ophelia. Vol. 17(1). 1978. P. 95-120.

Petraitis PS; Dudgeon S. Detection of alternative stable states in marine communities. *J. Exp. Mar. Biol. Ecol*. 2004, 300, 343-371.

Pianka E.R. On *r*- and *K*-selection. *Am. Nat*. 1970, 104, 592-597.

Plotkin AS; Railkin AI; Gerasimova EA; Pimenov AYu; Sipenkova TM. Subtidal underwater rock communities of the White Sea: structure and interaction with bottom flow. *Russian Journal of Marine Biology*. 2005, 31, 335-343.

Rabotnov TA. Phytocenology. Moscow. *MSU*. 1992. (in Russian).

Rabotnov TA. On violents, patients and explerents. *Bull. Mosc. Soc. Natur. Biological Series*. 1993, 98, 119-124. (in Russian).

Rajbanshi R; Pederson J. Competition among invading ascidians and a native mussel. *J. Exp. Mar. Biol. Ecol*. 2007, 342, 163-165.

Railkin AI. Marine Biofouling − Colonization Processes and Defenses. New York. CRC Press. 2004.

Ramenski LG. On the principal points, main concepts and terms using in plant ecology. *Sovetskaja Botanika.* 1935, 4, 25-42. (in Russian).

Rao KS; Balaji M. Observations on the development of test block communities at an Indian harbor. In: Thompson MF. et al. ads. Recent developments in biofouling control. New Dehli (India). *Oxford and IBH.* 1994, 75-96.

Redikortsev VV. Tunicata. Fauna of Russia and adjacent countries. *Bulletin of the Imperial Academy of Sciences.* 1916, 1. (in Russian).

Reish DJ. Studies on *Mytilus edulis* community in Alamitos Bay, California: I. Development and destruction of the community. *Veliger.* 1964a, 6, 124-131.

Reish DJ. Studies on *Mytilus edulis* community in Alamitos Bay, California: II Population varuiation and discussion of the associated organisms. *Veliger.* 1964b, 6, 202-207.

Riggio S. The fouling settlements on artificial substrata in the harbour of Palermo (Sicily) in the years 1973-1975. *Quad. Lab. Tecnol. Pesca,* Ancona. 1979, 2, 207-242.

Rocha RM. Replacement of the compound ascidian species in a southeastern Brazilian fouling community. *Boletim do Instituto Oceanografico Sao Paulo.* 1991, 39, 141-153.

Romanovski YuE. Life cycle strategies: a synthesis of empirical and theoretical approaches. Journal of General Biology 1998, 59. N6, 565-585. (in Russian).

Rudyakova NA. Fouling in the North-West Pacific. Moscow. *Nauka.* 1981. (in Russian).

Russell BC. The development and dynamics of a small artificial reef community. *Helgolland Wiss. Meeresuntersuch.* 1975, 27, 298-312.

Saenger P; Stephenson W; Moverley J. The subtidal fouling organisms of the Calliope River and Auckland Creek, *Central Queensland. Mem. Queensl. Mus.* 1979, 19, 399-412.

Scheer BT. The development of marine fouling communities. *Biol. Bull.* 1945, 89, 103-122.

Seed R. Patterns of biodiversity in the macro-invertebrate fauna associated with mussel patches on rocky shore. *J. Mar. Biol. Ass. U.K.* 1996, 76, 203-210.

Sell D. Marine fouling. *Proc. Royal Soc. Edinburgh.* 1992, 100B, 169-184.

Sejr MK; Sand MK; Jensen KT; Petersen JK; Christensen PB; Rysgaard S. Growth and production of *Hiatella arctica* (Bivalvia) in a high-Arctic fjord (Young Sound, Northeast Greenland). *Mar. Ecol. Progr. Ser.* 2002, 244, 163-169.

Shelford VE. Geographic extent and succession in Pacific North American intertidal (Balanus) communities. *Publ. Puget Sound Biol. Sta.* 1930, 89, 217-224.

Shilin MB; Oshurkov VV. Vertical distribution and some characteristics of the settlement of fouling planktonic larvae in the Kandalaksha Gulf of the White Sea. In: Scarlato O.A. ed. Ecology of the fouling in the White Sea. Leingrad. *Zoological Institute AS USSR*. 1985, 60-66. (in Russian).

Sirenko BI; Kunin BL; Oshurkov VV; Kataeva TK; Babkov AI; Golikov AN; Khlebovich VV; Kulakowski EE. Succession of fouling communities on artificial substrata in the White Sea. In: Spatial distribution and ecology of coastal byocenoses. Leningrad. *Nauka*. 1978, 10-13. (in Russian).

Stoecker D. Resistance of a tunicate to fouling. *Biol. Bull. Mar. Biol. Lab. Woods Hole*. 1978, 155, 615-626.

Stoecker D. Relationships between chemical defense and ecology in benthic ascidians. *Mar. Ecol. Prog. Ser.* 1980, 3, 257-265.

Stuart V; Klumpp DW. Evidence for food-resource partitioning by kelp-bed filter feeders. *Mar. Ecol. Progr. Ser.* 1984, 16, 27-37.

Sukhotin AA; Kulakowski EE. Growth and population dynamics in mussels (*Mytilus edulis* L.) cultured in the White Sea. Aquaculture. 1992, 101, 59-73.

Sutherland JP. Multiple stable points in natural communities. *American Naturalist*. 1974, 108, 859-873.

Sutherland JP. Life histories and the dynamics of fouling communities. In: The Ecology of Fouling Communities. Proc. US – USSR Workshop within the Program "Biological Productivity and Biochemistry of the World's Oceans". Part I. Beaufort. *North Caroline*. 1975, 137-151.

Sutherland JP. Functional roles of *Schizoporella* and *Styela* in the fouling community at Beaufort, North Carolina. *Ecology* 1978, 59, 257-264.

Sutherland JP. The fouling community at Beaufort, North Carolina- a study in stability. *Am. Nat.* 1981, 118, 499-519.

Sutherland JP; Karlson RH. Development and stability of the fouling community at Beaufort, North Carolina. Ecological Monographs. 1977, 47, 425-446.

Svane I; Setyobudiandi I. Diversity of associated fauna in beds of the blue mussel Mytilus edulis L.: effects of location, patch size, and position within a patch. *Ophelia*. 1996, 45, 39-53.

Thomassen S; Riisgård HU. Growth and energetics of the sponge *Halichondria panicea*. *J. Mar. Ecol. Progr. Ser.* 1995, 128, 239-246.

Todd CD; Turner SJ. Ecology of intertidal and sublittoral cryptic epifaunal assemblages. 3. Assemblage structure and the solitary/colonial dichotomy. *Sci. Mar.* 1989, 53, 397-403.

Tsuchia M; Nishihira M. Islands of *Mytilus edulis* as a habitat for small intertidal animals: effect of *Mytilus* age structure on the special composition of the associated fauna and community organization. *Mar. Ecol. Progr. Ser.* 1986, 31, 171-178.

Turner SJ; Todd CD. The early development of epifaunal assemblages on artificial substrata at two intertidal sites on an exposed rocky shore in St. Andrews Bay, N.E. Scotland. *J. Exp. Mar. Biol. Ecol.* 1993, 166, 251-272.

Turpaeva E.P. Biological model of fouling community. Moscow. *Institute of Oceanology AS USSR.* 1987. (in Russian)

Underwood AJ; Anderson MJ. Seasonal and temporal aspects of recruitment and succession in an intertidal estuarine fouling assemblage. *J. Mar. Biol. Ass. UK.* 1994, 64, 563-584.

Vance RR. Ecological succession and the climax community on a marine subtidal rock wall. *Mar. Ecol. Progr. Ser.* 1988, 48, 125-136.

Vasilevich VI. Types of plant strategies and phytocenotypes. *Journal of General Biology.* 1987, 48, 368-375. (in Russian).

Venkat K; Anil AC; Khandeparker DC; Mokashe SS. Ecology of ascidians in the macrofouling community of New Mangalore port. *Indian J. Mar. Sci.* 1995, 24, 41-43.

Wendt PH; Knott DM; Van Dolah RF. Community structure of the sessile biota on five artificial reefs of different ages. Bull. Mar. Sci. 1989, 44, 1106-1122.

Westinga E; Hoetjes PC. The intrasponge fauna of *Spheciospongia resparia* (Porifera, Demospongia) at Curacao and Bonaire. *Mar. Biol.* 1981, 62, 139-150.

Wahl M; Banaigs B. Marine epibiosis. III. Possible antifouling defense adaptations in *Polysyncraton lacazei* (Giard) (Didemnidae, Ascidiacea). *J. Exp. Mar. Biol. Ecol.* 1991, 145, 49-63.

Whomersley P; Picken GB. Long-term dynamics of fouling communities found on offshore installations in the North Sea. *J. Mar. Biol. Ass. U.K.* 2003, 83, 897-901.

Yakovis EL; Artemieva AV; Fokin MV; Varfolomeeva MA; Shunatova NN. Effect of habitat architecture on mobile benthic macrofauna associated with patches of barnacles and ascidians. *Mar. Ecol. Progr. Ser.* 2007, 348, 117-124.

Yereskovski AV. Some characteristics of inhabitation and distribution of sponges in the intertidal zone of East Murman. *Zoologicheski zhurnal.* 1994, 73, 5-17. (in Russian).

Yereskovski AV; Semenova MM. Species composition and certain ecological characteristics of epibiontic sponges in the White Sea. In: Sponges and cnidarians. Current status and prospects for future research. Leningrad. *Zoological Institute AS USSR.* 1988, 12-16. (in Russian).

Zajac RN; Whitlatch RB; Osman RW. Effects of inter-specific density and food supply on survivorship and growth of newly settled benthos. *Mar. Ecol. Progr. Ser.* 1989, 56, 127-132.

Zevina GB. Fouling in the White Sea. *Proceedings of Institute of Oceanology AS USSR.* 1963, 70, 52-71.

INDEX

A

abiotic, 8, 9, 18, 42, 45
acetate, 23
adaptations, 50, 65
adult, 22, 24, 25, 30
age, 23, 25, 64
algae, 14, 30, 33, 38
alternative, 1, 5, 49, 61
anatomy, 58
animals, 9, 21, 22, 23, 24, 25, 26, 29, 31, 34, 35, 37, 39, 40, 45, 57, 64
Antarctic, 30, 50
application, 23, 45
aquaculture, 55
Arctic, 9, 30, 63
argument, 44
asian, 60
assessment, 44
attachment, 34, 41
authors, 4, 8, 23, 26, 28

B

bacteria, 49
basal layer, 27
behavioral aspects, 50
biodiversity, 63

biomass, 8, 14, 28
biomass growth, 8
biota, 65
biotic, 18, 42, 45, 53
biotic factor, 42
bivalve, 39
Black Sea, 51
body shape, 22
body size, 26
body weight, 35
Brazilian, 62

C

cavities, 34
chemicals, 39
classification, 7, 8
clustering, 37
colonization, 50, 54
communication, 26, 27, 30
competition, xi, 7, 31, 34, 35, 51, 60, 61
competitive advantage, 42
competitor, 18, 19, 33, 43, 61
complexity, 49
composition, 9, 45, 51, 64, 66
concentration, 26
consumers, 8
continental shelf, 54
control, 61, 62

covering, 33
CRC, 61
criticism, 44
cues, 37
cultivation, 23, 24, 25
cultivation conditions, 25
culture, 22, 55, 58
cycles, 61
cytotoxic, 39

D

death, 3
defense, 63, 65
degradation, 22
density, 7, 8, 28, 66
destruction, 4, 62
dichotomy, 64
distribution, 28, 55, 58, 59, 63, 66
diversity, 1, 4, 32, 55
dominance, 3
dry matter, 40
drying, 14
durability, 39
duration, 2

E

ecological, 7, 8, 18, 44, 50, 51, 53, 59, 61, 63, 66
ecosystems, 44
environment, 8, 28, 29, 31, 32, 37, 39, 44, 45, 53
environmental conditions, 8, 9, 18
environmental factors, 18, 31, 41
epibenthic communities, xi, 1, 34, 35, 37, 44, 49, 60
estimating, 5
estuarine, 27, 52, 65
evolution, 51
exposure, 17, 39, 60

F

family, 14, 29
Far East, 53
fauna, 18, 32, 33, 50, 53, 58, 59, 60, 63, 64, 65
feeding, 35
fertility, 18, 26, 27, 43
film, 13
filter feeders, 63
filtration, 37, 57
floating, 53
flora, 32, 60
flora and fauna, 32
flow, 14, 61
fluctuations, 28, 42, 59
food, 18, 35, 45, 63, 66
freezing, 10
fresh water, 28, 29

G

groups, 27, 34
growth, 7, 8, 18, 21, 22, 24, 25, 26, 33, 38, 50, 56, 59, 66
growth rate, 18, 21, 25, 26, 38, 56

H

habitat, 23, 26, 59, 64, 65
harbour, 62
heart, 57
heart rate, 57
heating, 13, 28, 29, 50
height, 21, 22
hydroid, 38
hydrological, 9, 13, 22, 29
hypothesis, ix, 56

I

ice, 3, 10
immersion, 29, 51
Indian, 30, 62, 65
Indian Ocean, 30
industry, 55
inhibition, 3
interaction, 36, 61
invertebrates, xi, 45, 51, 52, 54, 59
Investigations, 50
irrigation, 33
island, 49

J

juveniles, 24, 25, 38

K

kelp, 14, 63

L

land, 44
larvae, 17, 26, 27, 30, 37, 38, 39, 43, 56, 59, 63
layered architecture, 32
life cycle, 22, 54
life span, 2, 18, 23, 24, 25, 38, 39, 45
life style, 9, 31, 32, 38
line, 18
linear, ix, 4, 25, 35
linear function, 35
living conditions, 7, 31, 33
location, 19, 64
long period, 28
longevity, 39
LSU, 54

M

mantle, 28
Markov model, 4
maternal, 22, 27
measures, 22
Mediterranean, 49
melting, 10
metabolites, 39
metamorphosis, 37, 39, 53, 56
microbial, 13, 52, 56
model, xi, 3, 4, 7, 9, 40, 44, 45, 65
mollusks, ix, 17, 23, 26, 28, 33, 37, 45, 50
monsoon, 3, 29
morphological, 40
mortality, 7, 8, 29
mosaic, 33, 43
multidisciplinary, 57
mussels, xi, 3, 14, 21, 25, 26, 27, 28, 31, 32, 33, 34, 37, 38, 50, 57, 64

N

natural, xi, 23, 52, 64
necrosis, 39
network, ix, 4
NOAA, 54
normal, 30
North America, 63
North Atlantic, 23, 32
Northeast, 63
nutrient, 57

O

observations, 2, 22, 23, 24, 27, 29, 30, 33, 37, 45
offshore, 65
oocytes, 26
open space, 34
order, 23, 27

organism, ix, 3, 4, 8, 17, 34, 43, 50
oysters, xi, 3, 52

P

Pacific, 30, 53, 62, 63
parameter, 18, 21, 22, 24, 25, 27, 31, 33, 35, 36
particles, 35, 59
patients, 61
periodic, 2, 7, 14, 17, 29, 39, 42
personal communication, 26, 27, 30
physical factors, 4
physical properties, 14
physiological, 50, 43
plankton, 26, 59
plants, 7, 8, 9, 40, 44, 53
plastic, 24
plasticity, 27
play, 2
population, 7, 8, 61, 64
population density, 7
population growth, 7
predation, 3, 49, 51, 61
prediction, 5, 54
pressure, 7, 8
production, 40, 56, 63
property, iv, 4, 35, 36
protection, 38, 39
pumping, 35, 36, 37, 57

R

rainfall, 54
range, xi, 28, 29, 30, 35, 42
RAS, 50, 55, 56, 57, 58, 59, 60
reason, xi, 2, 14, 31, 33, 35, 38, 42
recurrence, 1
reef, 49, 52, 62
reflection, 2
region, 10, 14
regular, 4

relationship, 25, 35, 61
relevance, 53
reproduction, 7, 27, 50
resistance, xi, 18, 36, 44
resources, 7, 8, 18, 33, 35, 36, 60
rocky, 52, 58, 63, 64
Russian, 7, 47, 49, 50, 51, 52, 53, 54, 55, 56, 57, 58, 59, 60, 61, 62, 63, 65, 66
Russian literature, 7

S

salinity, 9, 10, 13, 18, 28, 29, 30, 42, 50, 56, 57
saltwater, 55
sampling, 45
sand, 55
scarce resources, 36
scores, 19
search, 4
seasonality, 2, 3, 51
sedentary, 9, 14, 17, 33, 35, 37, 38, 39, 40, 42, 43, 44
sedimentation, 35
seizure, 26
series, ii, 2, 4, 35, 45
settlements, 3, 18, 21, 23, 28, 31, 32, 34, 37, 38, 39, 58, 62
shape, 22, 28
shelter, 18
shores, 25, 58
short period, 43
short-term, 2
sites, 64
solitary ascidians, ix, 3, 32
space, 19, 34, 40, 52, 61
spatial, 55, 57
spawning, 22, 24, 26
species, ix, 2, 3, 4, 5, 7, 8, 9, 17, 18, 19, 22, 23, 24, 25, 26, 27, 28, 30, 31, 32, 33, 34, 35, 36, 37, 38, 39, 40, 41, 42, 43, 44, 45, 51, 55, 62

specificity, 50
spectrum, 35, 45
sponge, ix, 3, 13, 14, 17, 22, 25, 30, 33, 39, 40, 43, 49, 50, 52, 53, 54, 56, 64, 66
stability, 52, 64
stable states, 61
stages, ix, 2, 43
stochastic, 1, 2, 4
stochastic model, 4
storms, 3
strategies, ix, 7, 8, 9, 17, 18, 19, 27, 40, 42, 43, 44, 45, 53, 56, 62, 65
strength, 18, 34, 57
stress, 8, 43
structural characteristics, 40
structuring, 55
substances, 39
substitution, 4
substrates, xi, 3, 17, 22, 26, 28, 29, 34, 37, 38, 55
succession, ix, 1, 2, 3, 4, 5, 9, 14, 32, 44, 51, 52, 54, 55, 58, 60, 63, 65
summer, 10, 13, 23, 24, 28, 29, 33, 50
supply, 66
surface water, 10, 13, 28, 29
survival, ix, 29, 30, 56
symbiosis, 49
synthesis, 62

T

temperature, 10, 18, 28, 29, 30, 42
temporal, 53, 65
tension, 35
tides, 14
time, xi, 18, 22, 28, 29, 38, 39, 42, 43
time frame, 22
timing, 60
tissue, 39
tolerance, ix, 7, 8, 18, 19, 25, 27, 28, 29, 30, 33, 36, 39, 41, 42, 43, 52, 56, 57
toxic, 39, 49
traits, ix, 42

V

values, 8, 36
variability, 2
variation, 58
vector, 8, 40
vegetation, 52, 53
violent, 7

W

water, 9, 10, 13, 15, 17, 28, 29, 30, 31, 35, 42, 44, 56
winter, 10, 22, 28